GREENER ENERGY SYSTEMS

GREENER ENERGY SYSTEMS

Energy Production Technologies with Minimum Environmental Impact

Eric Jeffs

CRC Press is an imprint of the
Taylor & Francis Group, an **informa** business

CRC Press
Taylor & Francis Group
6000 Broken Sound Parkway NW, Suite 300
Boca Raton, FL 33487-2742

Printed in the United States of America on acid-free paper
Version Date: 20120227

International Standard Book Number: 978-1-4398-9904-5 (Hardback)

Library of Congress Cataloging-in-Publication Data

Jeffs, Eric J.
Greener energy systems : energy production technologies with minimum environmental impact / Eric Jeffs.
p. cm.
Includes bibliographical references and index.
ISBN 978-1-4398-9904-5 (hbk. : alk. paper)
1. Electric power systems--Environmental aspects. 2. Electric power production--Environmental aspects. 3. Renewable energy sources. 4. Sustainable engineering. I. Title. II. Title: Energy production technologies with minimum environmental impact.

TK1001.J44 2012
621.31028'6--dc23 2012001487

Visit the Taylor & Francis Web site at
http://www.taylorandfrancis.com

and the CRC Press Web site at
http://www.crcpress.com

CONTENTS

ACKNOWLEDGEMENTS

I would like to thank the many people in the electricity supply industries and their suppliers who have helped me over the years, providing information and arranging site visits around the world. In particular I thank the three people who have reviewed my manuscript and for their valuable comments.

Dick Foster Pegg was born in the United Kingdom and served an apprenticeship at Rolls-Royce at the time of their early gas turbine models, before emigrating to the United States where he worked for Bechtel Corporation on power plant design. Later he joined Westinghouse to work on gas turbine research and development in Pennsylvania.

Louis Codogno was Managing Director of the Energy Division of Cockerill Mechanical Industries until his retirement in 2005. I thank him in particular for not only arranging site visits but also introducing me to many of his clients which gave me an insight into the energy policies of those countries, notably in the Middle and Far East.

Bo Svensson was born in Sweden and spent his entire career at one of the country's oldest engineering companies, Stal-Laval Turbin AB, now Siemens' Industrial Turbomachinery Division, first as a project engineer, and later as the Sales and Marketing Manager for gas turbines. Since his retirement in 2001, he has been a regular contributor to Diesel & Gas Turbine Worldwide and Diesel Progress.

FOREWORD

During the winter of 2008-2009 I wrote Green Energy, a book which looked at the energy technologies that could meet the growing energy demands of a world which was then obsessed with the concept of Global Warming and Climate Change. It was published in November 2009, a few weeks before the International Conference on Climate Change in Copenhagen, which was intended to plot a path towards setting up a global standard of carbon dioxide emissions.

In the event the conference was judged a failure and it was ironic that on the last day of the meeting snow fell in Denmark, and over much of Northern Europe, which was so unusual that, to the general public, it somewhat gave the lie to global warming. December 2010 again saw heavier snowfall which, although it did not fall in London on Christmas day, did not really start to thaw for another two or three days after that.

If global warming is genuine, is it not a natural climate change and not the result of emissions which technology has over the last 100 years progressively reduced in every process that produces them? This has brought more efficient engines, electrification of railways, the introduction of nuclear energy, and a steady decline in the production and use of coal, not least in the home where natural gas is now the principal fuel for heating and cooking in Europe and North America.

How then do we generate enough electricity, efficiently and with mininum emissions to meet more of our future energy needs across the world? Given that there is a great disparity between per capita electricity consumption in Canada, in Europe, and in Africa, where large swathes of sub-Saharan regions have no access to electricity supply at all, it is questionable that this situation can be changed easily. The millions of Asians and Africans with no access to electricity supply are one of the great sources of pollution in the world, dependent as they are on wood and dried animal dung for their basic daily energy needs.

Then in the developed world there is a large inventory of coal-fired plants still operating after more than sixty years at an efficiency of less than 34%. New regulations on emissions may force many of these to close in the United States, as similarly the European Large Combustion Plant Directive will force many coal- and oil-fired plants built before 1987 to close at the end of 2015 across the 27 countries of the European Union

The first act in the light of all this is to maintain security of electricity supply over the next ten years which means having enough operating plant to meet demand in 2016 and enough under construction to meet ongoing demand in following years. Demand is sure to increase as more electrification of transport occurs along with more combined heat and power schemes for industry and for district heating. For as important as is security of supply it is equally so the efficiency with which electricity is produced.

Efficiency has almost doubled in the last 70 years. In 1945 a typical coal-fired plant of the day would have an efficiency of about 30%. Today a larger plant with a supercritical steam cycle can achieve 45%. Meanwhile the largest gas-fired combined cycles typically operate at about 58%, while the latest design has achieved 60.75% at a plant in southern Germany.

To go any higher it is desirable to use some of the waste heat from power generation. Combined heat and power has expanded in the last twenty years, but there are many applications which could still be exploited, not least in the expansion of district heating and cooling in many more cities.

1
Present energy demand

Two extremely cold winters in Europe in successive years have done more to increase public cynicism about global warming. Yet there is now evidence of increased carbon dioxide in the atmosphere. Is it because of industrial and transport emissions, which technology has steadily reduced over the years, or rather because the world population has trebled to over six billion since the end of the Second World War, and could reach nine billion by 2050.

Against this background energy demand is bound to increase and it must be done in a way that does not impinge on the other needs of a growing population: the production of more food, potable water, and places to live. It must also be produced in a way which does not make an excessive impact on the environment.

We cannot reduce the population by war or disease but we can reduce energy consumption with new technology and produce it more efficiently. Given our existing technologies that effectively rules out coal as a suitable fuel, because as we have already seen with flue gas desulphurization, any add-on system to make the plant more environmentally acceptable will add to the auxiliary load so that less electricity will be produced, and more significantly push up the cost to the consumer. Provision of carbon capture and storage is a condition required of any new coal-fired plant. But the technology is in its infancy and for a given quantity of coal less electricity will be produced at a far greater cost.

Other technologies such as nuclear and combined cycle have developed on an opposite track. The developers of nuclear reactors seek to make them simpler to construct and therefore maintain their low cost of electricity production. Also smaller reactors are now being designed which can broaden the application of nuclear energy to shipping and combined heat and power for industry. Indeed one may ask, given the experience of small reactors of less than 100 MW output in nuclear submarines, aircraft carriers, and ice-breakers over the last fifty years, why these applications have not been tried before. One has only to point a finger at the Green movement to get the answer.

In fact Green Activists have until now had a field day. They came into being after the 1973 oil crisis and although their first concerns were about public health issues, such as lead in gasoline, they very quickly focused on American applications to build more nuclear power plants, particularly the fast breeder reactors which used a plutonium fuel cycle.

The American aim, as in the other industrial countries, was to take oil out of power generation. Forty years later, oil contributes less than 3% to the total electricity generated in the country and in that they have succeeded. Oil is mainly used as a reserve fuel for the gas-fired steam and combined cycle plants. But it was not until George W. Bush arrived in the White House in 2000 that the revival of nuclear power began in earnest.

The late sixties and early seventies were the middle years of the cold war. Relatively fresh in the public mind was the stand-off between the United States and the Soviet Union over a plan to station ballistic missiles in Cuba. Green anti-nuclear views fell on fertile ground because many people, particularly on the political left, were of the opinion that we had only narrowly avoided nuclear war.

There are several Green parties in Europe, in some cases holding the balance of power as a minority group in a coalition government. Indeed when they actually got into power in Germany in 1997, they immediately wanted to shut down all the nuclear power stations which then supplied 33% of the country's electricity.

Besides raising public fears about nuclear energy, they have also attacked genetically modified (GM) crops as a source of "Frankenstein food," which would be unsafe to eat, and which would kill off biodiversity. In Europe, trial plantings of GM crops have been trashed by Green Activists in the fields where they were growing. Yet farmers in Africa and India have happily planted and harvested GM crops in the knowledge that they are of better quality with less pest damage, and therefore give them a higher income.

Anybody who has visited the United States in the last five years will probably have unknowingly eaten something which has been prepared from a genetically modified plant, or the meat of a cloned animal, and lived to tell the tale. But not in Europe, where GM crops and cloned animals are subjects for study and not for human consumption. When a cloned animal is killed the carcass must be burned. Yet in the autumn of 2010 there was a huge outcry because meat from a cloned animal had inadvertently got into the UK food chain.

The Green influence in European Governments has been profound. Protests against capital projects have seriously delayed, if not cancelled, their implementation. Their contribution to energy policy has been based on opposition to nuclear energy and the advocacy of renewable energy systems, except traditional Hydro Power, which they vehemently oppose.

The result has been the creation of the wind power industry which might never have happened without them, and their influence on governments particularly in Europe, who have created energy policies which specify a percentage of renewables, principally wind power, to be included in the plant mix.

Combined cycle goes for higher gas turbine performance and higher efficiency. As an electricity producer only, efficiency has risen to 60 %, as compared with 45% for a large coal-fired supercritical steam plant. But the big gas turbine applications have been to industrial combined heat and power schemes which started in the United States with the Public Utility Regulatory Policies Act (PURPA) of 1979, and was directed specifically to industrial power generation.

Privatization of the British electricity supply system in 1990 and subsequent deregulation of electricity supply in other countries saw combined heat and power expand around the world. Deregulation also gave the impetus to deployment of renewable energy systems: first a plethora of wind farms, which was followed by the development of photovoltaic semiconductors which have found application in solar energy schemes for offices and some homes; and feed water heaters for some combined cycles.

Other renewables are hydro power but not the large hydro plants of today but rather redevelopment of the small hydro sites from 80 to 100 years ago which were built for a specific job wherever suitable water supply was found. Many were abandoned as large thermal power plants were built to deal with a growing electricity demand after the Second World War. Such hydro schemes would have then been almost 50 years old with outputs between 100 kW and 10 MW.

Biomass is the other renewable which is judged to be green because burning it simply releases the carbon dioxide absorbed in its growth. But it doesn't mean cutting down trees to fuel a wood-fired power station. There are some tree species which are fast growing and can be pollarded; selected branches are cut and, in typically three years, new growth can be harvested again. Then there is cropping for biomass with specialist crops such as miscanthus (elephant grass) which is planted as rhyzomes that sprout every year, and can be harvested annually in spring using a combine harvester.

The one biomass success story which has been running for over forty years is in Brazil, the world's leading sugar producer, which in the early 1970s was an oil importer. The government after the first oil crisis authorised the planting of more sugar which could be distilled to make alcohol as a fuel to blend with gasoline. By the end of that decade cars throughout Brazil were running on a gasoline/alcohol mixture containing about 5% alcohol.

Biomass however has political and economic problems. A growing world population puts many demands on land use, and not only for agriculture. Every hectare used to grow biomass fuel is one less on which to grow food crops. So the more biomass crops are grown the less food crops and so the price of basic food stuffs increases. Biomass power plants alone are small, and are mainly using scrap wood from paper mills and sawmills. But some old coal-fired power stations are regularly burning biomass as up to 10% of their fuel input.

There are a number of issues behind energy production in the future. Where can we find it, and in what environment, and what are the risks attached to developing it?

First, how much electricity do we need? About 6000 kWh/year

is roughly the average individual consumption in Europe. The typical electricity use in European homes is with a refrigerator and freezer to store food; a washing machine and tumbler drier to clean and dry clothes; other small electrical appliances such as a vacuum cleaner, electric iron; several radios, and television sets, and one or more computers.

Cooking, if not by gas would be with an electric cooker; and in any case there might also now be a microwave cooker. Water heating is generally from the same gas or oil-fired boiler that supplies space heating but there will be an immersion heater in the tank to heat water during the summer months.

Since we are entering an 11-year cyclic period of greater solar activity we may experience a hitherto benign climate change. But perhaps more important is the future price of oil. If it becomes too expensive will people switch more to diesel or electric cars, which in one case are more fuel-efficient and in the other have much lower fuel costs? If they do, then the Greens Activists are waiting for them.

In the UK some local authorities are looking at charging more for diesel cars in permanently paid long-term car parks, because of the particulate matter in the exhaust. This might have been true fifty years ago when it was common to see diesel trucks climbing a hill with great clouds of black smoke issuing from the exhaust. The huge public protest at the time resulted in threats to take such vehicles off the road, which very quickly led to improvements in the design and maintenance of diesel engines.

The other energy saving protest was that fluorescent lights were less bright that the traditional incandescent bulbs, which are no longer in production. A 20 W fluorescent lamp can produce as much light as a 100 W incandescent bulb, and the only way to get a bulb now is from a shop that has not completely cleared its old stock; if such can be found.

So there are two energy saving measures which have been a matter of personal choice, and one of which, at least, is now permanent and offers to produce a considerable saving in electricity demand. But still some of the greatest energy loss is from office buildings, and homes. Single glazing and no loft insulation in older houses is one of the reasons why. For new houses, current building standards require them to be heavily insulated in the wall cavities and roof space, and also to have double-glazed windows.

All these measures can be back fitted in existing houses, some more easily than others, as many people have already done. The double-glazing units are supplied as prefabricated modules to be fitted in UPVC high strength plastic frames which require no painting. Roof insulation is available in 400 mm wide by 10 m long rolls which fit in between the rafters. As more new houses and converted older houses account for a greater proportion of the housing stock the demand for heat is likely to reduce.

Few houses have been built with electric heating, but many old houses with coal fires in selected rooms have converted to electric heating. Forty years ago as the first nuclear power stations came into operation, to counter their poor load-following capability, night storage heating was introduced.

The householder, in each room, had to install heaters consisting of a ceramic block of high thermal capacity in a casing forming a duct around it. The blocks are heated during a defined period at night when electricity is charged at a special low price per kWh. The householder can control the rate of charge and of discharge, for each heater.

The room is effectively heated by convection; the heater is designed so that it stands about 10 cm off the floor, which allows cold air to flow in at the bottom which is heated as it passes over the core and out through a grille at the top.

In the UK the charging period is seven hours from midnight to 07.00 a.m. GMT, all year round. During this period the price per kWh of electricity is one third of the price at other times. People with this tariff also heat water and use programmable appliances such as a washing machine or a dishwasher, to take advantage of the lower rates. The modern machines have a single cold water feed which they heat to the required temperature for the program.

The Greens give an impression that much of the present day lifestyle will have to be given up to counter global warming and avoid having to build more nuclear power stations. But the fanatics are beginning to fall out among themselves, some of them realise the importance of electricity in running society and the value of nuclear plants as having no emissions in operation.

In any case, in spite of its removal from a large part of electricity production, oil is still the only fuel available for transport on roads and in the air. We must therefore move energy consumption to other fuels than oil, and in the home to electricity. To drastically reduce energy consumption would not be good for public health even if it meant we had to walk more often.

The refrigeration of foodstuffs has been a major factor in the improvement of health over the years. Until sixty years ago most women walked to nearby shops almost every day to buy food for their family. Once ownership of refrigerators became widespread it was possible to store food, particularly meat, milk and vegetables in conditions which protected them from changes in temperature that might have made them sour or otherwise unfit to eat.

As daily food shopping reduced, and car ownership became more widespread, the big grocery companies built supermarkets which include the traditional trades of greengrocer and butcher and have expanded into clothing and electrical goods. Other facilities included are Automatic Teller Machines so the customers can withdraw money, and filling stations attached to the car parks where often gasoline is cheaper than outside at the regular filling stations. A round trip of about 10 to 20 km to the supermarket once a week is now normal for most families.

A large supermarket has several aisles of freezer cabinets for particularly meat, both fresh and manufactured meat products, packed in standard quantities, and frozen vegetables and all to be sold, if not eaten, before a certain date printed on the packet. The whole shop would be air-conditioned to remove the heat from the refrigerators and freezers and provide a comfortable environment for people to walk around and make their purchases. So each supermarket represents a large electricity load 24 hours a day.

In many countries oil has been taken out of the power generation market except for use as a back-up fuel for combined cycles and other gas-fired power plant. This has been relatively easy to do, and similarly the electrification of some railways has lowered the demand for diesel fuel for locomotives. There are still many long distance routes which are not electrified, particularly in the United States and Canada, and the United Kingdom. Diesel power would be retained for marshalling yards and at docksides and on small branch lines where it would be uneconomic and impractical to string overhead conductor wires.

High speed trains are now commonplace across Europe and some services, notably Eurostar through the Channel Tunnel, have taken some business from the airlines. There have been some high-speed trains introduced in the Far East, notably in Korea, but in Japan the Shinkansen system predates all the others. The Shinkansen was built to provide a higher speed service across Japan to all four main islands because the original railway network was built on a metre gauge. Shinkansen is therefore a standard gauge system as are all the European high-speed services.

This summary of present energy use is met with the present electricity generation system. But a much larger switch to electricity in transport and space heating in rural areas would require a large program of power station construction and grid connexions.

Given that to build a power plant first involves legal issues of deciding where to build it and obtaining a construction permit; and assuming that all consents have been obtained for construction and connexion to the grid, a 500 MW combined cycle block could be up and running in about 30 months. In any event more combined cycles will be needed because with their flexible operation and rapid starting capability they are important as back-up for renewables such as wind and solar which have variable outputs.

A 1200 MW nuclear plant with the currently available new reactor designs could also be up and running in five years. Again reactor design has produced standardized units from all the leading manufacturers in North America, Europe, Russia and the Far East, and all of which are designed for a service life of sixty years.

Not only must more generating capacity be built, and also to replace old plants, but intelligent grids must be deployed. An intelligent grid is one which can look where the greatest energy demand is occurring and bring on under-utilised plant to meet the load whenever it is required.

The intermittent nature of wind power created the demand for the intelligent grid and a widespread conversion to electric cars will also need it. Say half the houses in a street of 100 each had one electric car which they charged every night that could be around 1500 kWh taken over seven hours. But the use of the cars will vary and the amount drawn every night might vary considerably from day to day.

How fast will it all happen? In Europe at least the ambition is to have reduced greenhouse gas emissions by 80% up to 2050. But current projections of population growth suggest that food production must increase by 70% over the same period. Surely this rules out biomass energy on a large scale and other renewables which may interfere with farm land (on-shore wind) or fishing (off shore wind and tidal stream systems).

This must mean a significant switch over to electric or hybrid cars and electric heating powered by nuclear plant, which generate electricity without emitting one gram of greenhouse gas. It cannot be achieved totally with renewable systems, although, particularly, photovoltaic cells can be an important accessory for an electric car to extend its range, and as a source of electricity supply to a large office building, and there are many opportunities for these applications around the world.

This is possible because of the conditions attached to energy production. Much more than in the past, national energy policies have been modified by international agreements. The first of these is that a given percentage of electricity production should be by renewable energy systems. The obvious system is wind, but wind farms occupy a large land or sea area to produce a relatively small amount of power. Fifty 5 MW offshore wind mills contain at least fifty times as much steel as is needed to make the pressure vessel of a 1000 MW nuclear power reactor. Think, too of the amount of energy required to produce and install the components of the wind farm.

Where legislation has aimed at individual consumers, systems serving individual buildings, such as ground-sourced heat pumps, or solar photovoltaic panels on the roof are possible. Heat pumps, of course have electric motors; and photovoltaic panels only operate in daylight hours. Nevertheless several large office buildings have installed solar panels to supply their computers and other equipment and lighting during the working day. Over a year in northern Europe this could provide up to 100 000 kWh/year, and much more in sunnier climates.

Small rivers of steady flow can also have micro hydro sets of a few kW output. These have less environmental impact and in some countries, notably Canada, it has provided an opportunity to redevelop some of the very first sites, which were abandoned many years ago, and so increase the hydro power supply.

Utilities are offering feed in tariffs, which pay for electricity produced in homes and offices, which is not used. So the house with solar panels on the roof can operate electrical appliances during the day time with free electricity from the solar panels. If the house is unoccupied for all or part of the day, electricity fed back to the grid is paid for at the daytime tariff.

Renewable energy as a percentage of generation has not really performed as was hoped. Wind is intermittent and it is reported that during the snowy period of December 2010 the wind was very light and collectively British wind farms supplied at less than 10% of their rated output. The same situation could occur for a few days in a very hot summer.

The other policy relates to coal. Governments in the industrial world will allow coal-fired plants to be built only if they include a carbon capture and storage scheme, to remove carbon dioxide from the flue gases. While there are several schemes which are demonstrating the technology with a small percentage of the flue gases, there are no full scale schemes yet proposed to capture 90% or more of the carbon dioxide in the flue gases. All that is known is that studies have shown that carbon sequestration would seriously reduce the efficiency of the station and that the additional capital and fuel cost would add unacceptably to the cost per kWh produced.

So there is very little enthusiasm for coal-fired plant and with growing public acceptance of nuclear energy because of its low cost of production, its proven reliability and lack of emissions, that is surely the way to go. Inevitably we will be using more electricity to reduce emissions and the priorities of industry are in the development of more energy-efficient processes and products.

Not only this, unless there is a market for the recovered carbon dioxide, for example for enhanced oil recovery, the cost of the gas compressors, the pipeline, and the construction of the bore hole to the repository can only be passed on to the consumers as an additional cost for coal-fired energy.

Twenty years ago in Japan, Mitsubishi in collaboration with Kansai Electric, developed the KM carbon capture system for gas-fired boilers, which has been applied to industries with a demand for carbon dioxide. The manufacture of fertiliser from natural gas starts with the production of ammonia and then reacts with carbon dioxide to produce urea. Three installations in Asia and the Middle East have gas-fired combined heat and power schemes where the boilers are fitted with carbon dioxide recovery which then reacts with the excess ammonia to make more urea.

The first unit was installed by Mitsubishi Heavy Industries for Petronas Fertilisers at Kedah, Malaysia in October 1999 where it recovers 200t/d of carbon dioxide from flue gas. Six years later two 450 t/d units were installed for Indian Fertilizers at Aonla. Then, a year later a 400 t/d unit was installed at Abu Dhabi Fertilizer which went into operation in October 2006.

These units are all operating and have proven the concept of carbon dioxide recovery as a feedstock for urea. These are relatively small units of less than 1 million t/year of recovered gas, but they show that any industry with a requirement for carbon dioxide as a feedstock can recover it from their own gas-fired boiler plant.

The big electricity application in future will be to transport which is only just beginning. More railways can be electrified. Tramways are returning to some British cities to replace bus routes. But to what extent is there support for a return to public transport?

The increased car ownership in the twenty years after the Second World War was an expansion of personal transport which has continued to grow. So too has the price of gasoline and diesel fuel. If this trend continues what will be the result? Will there be a return to public transport or the introduction of electric or hybrid vehicles? This will reduce oil consumption but increase electricity consumption to charge batteries.

If the latter we will see the return of the high rates of growth of electricity demand of before 1970, which were largely due to the great changes in domestic life with many electrical appliances taking over jobs which were previously done by hand. Also there was the increased production of aluminium and later titanium which must be electrolytically refined.

But interestingly Europe is beginning to favour nuclear power again. In 2015 the Large Combustion Plant Directive will cause the shutdown of a large number of coal-fired plants particularly in Poland, Germany, and the United Kingdom. Sweden has abandoned phase-out and will build new plants to replace the current plants when they are retired. Finland has authorised a sixth nuclear plant to follow on from Olkiluoto 3 which is under construction and planned for service at the end of 2012. Belgium has extended the life of their seven nuclear power reactors to sixty years. The United Kingdom has plans for up to ten reactors and are currently evaluating reactor designs which should be complete by the end of 2011.

But the big worry now is whether the earthquake and tsunami in Japan, in March 2011, which precipitated a nuclear crisis at one of the four plants shut down by the earthquake, will be misinterpreted by countries which do not have fault lines running through them, have never experienced a tsunami, and have few if any nuclear power plants on coastal sites. The last thing we want now is for politicians to get frightened of nuclear power again and prevent or delay the deployment of the only green energy source that can provide us with the electricity we need for the rapidly growing future applications.

So it looks a though the Green anti-nuclear influence on energy policy is finally waning and Governments are seriously looking as to how future electricity supply can be had in an environmentally friendly manner without excessive increase in cost.

2
Energy production risks

Accidents can happen in industrial processes; spillages, releases of gas and explosions which by and large are contained within the premises. The energy industries are no exception, but the consequences can be vastly different. The grounding of a tanker and the subsequent release of crude oil into the water can kill fish and sea birds, and destroy the livelihood of a coastal community dependant on fishing.

Electricity supply can be cut for several reasons. A lightning strike on an overhead power line or a wind generator, the latter a quite common ocurrence, could knock out a circuit breaker, which might reconnect quickly. Another reason could be a circuit overload which might occur if part of a circuit is isolated for repair or maintenance so that the energy flow has to go the long way around and share other lines which might themselves become overloaded.

A third reason might be political. If there is not enough available capacity power cuts might be instigated to avoid overloading the system. In 1974 in the UK some 90% of electricity supply was generated by coal, and a strike by the Mining Unions led to power cuts on a daily basis to conserve dwindling coal supplies at the power stations.

In Japan, in March 2011, there was closure of four nuclear plants by a powerful earthquake and damage caused by the following tsunami that shut down 25% of the country's nuclear capacity, which cannot come back on until new transmission lines are built. The government therefore announced a programme of rolling power blackouts around the country.

These issues apart, the risks are mainly in fuel supply and plant operation: coal mining, nuclear fuel fabrication, and oil and gas production, of which probably the potentially most dangerous are fossil fuel production and transport.

Possibly the worst accident ever, because of when it happened was the wreck of the *Torrey Canyon.* The ship was one of the new super tankers which were increasingly coming into service. In fact it was the largest ship of any sort to have been wrecked at that time measuring 297 m long by 38.2 m beam and a draught of 20.4 m.

The vessel was carrying 119,000 tons of crude to the Milford Haven refinery in South Wales when it struck Pollard's Rock on the Seven Stones Reef, which lies between the Isles of Scilly and Land End, 12 km east of the islands. It would seem that the ship had gone south of Scilly when it should have gone North to Milford Haven and decided to cut up between the islands and the mainland. The accident happened on Saturday morning March 18, 1967.

However, the accident happened off a major holiday area. The then Prime Minster, Harold Wilson, had his holiday home on Scilly and many people took holidays in Cornwall and on the Islands. It was the weekend before Easter and Mr. Wilson had already gone down with his family during the parliamentary recess.

Various attempts to refloat the vessel were unsuccessful and on Easter Monday the tanker broke in two, which was how the entire cargo spilt out. As luck would have it, instead of the normally prevailing south westerlies, the wind was blowing from the north and so most of it went on the Brittany coast and the Channel Islands particularly Guernsey, where about 3000 tons of contaminated crude was recovered and dumped in a quarry which had been an ammunition store for the German forces during the Second World War. Many years later some 160,000 litres of the contaminated oil was sent to a reprocessing plant at Kingston upon Hull.

In the following years rains have washed more of the oil out of the subsoil and when a Guernsey official saw a bird struggling in the oil he took a picture and posted it on Facebook, which has since got the government to act on a final clean up. This has included putting oil consuming microbes into the residual oil in 2010 in the expectation that they would clean it up.

In 1967 there was less understanding of oil spills and how to prevent them, also there was no understanding of the properties

2.1 Scilly Isles, UK: *Torrey Canyon*, carrying 119000 tons of crude from Kuwait to Milford Haven refinery struck Seven Stones Reef and ten days later on Easter Monday 1967. broke up and sank. (Photo courtesy of Daily Mail)

of the detergents that had been used to attack oil spillages up to then. They were seen to dissolve the oil and cause it to sink but they were later found to be highly toxic and would have killed marine life that came into contact with it. That would have been in addition to approximately 15000 birds killed as well as seals and other marine life along the south coast of England.

This was a first of a kind accident and there was a lot to be learned from it. One consequence of the *Torrey Canyon* disaster was the introduction of law on transport of crude and clean up after spillage. It was less than three years since the first gas was discovered in the North Sea and three before the first oil discoveries further north. Forty-five years later the sea has largely recovered and the wreck of the *Torrey Canyon* is one of the popular dives for people holidaying on Scilly.

July 1988 saw the destruction of the *Piper Alpha* platform in the North Sea it killed 167 of the 229 people on board and resulted in the highest total insurance claim up to that time of $1.4 billion. Occidental's *Piper Alpha* was one of three platforms in the area and the hub for oil transport to the mainland. The *Claymore* and *Tartan* platforms, each about 20 km distant, pumped their oil to

Piper Alpha and into the main pipeline ashore and these platforms certainly fed the fire until they realised what had happened.

Piper Alpha had started production in 1976 as an oil producer later it was converted to produce gas condensate and natural gas which compromised the original safety concept. On the morning of July 6, 1988, routine maintenance started on one of two condensate pumps, and in particular the pressure safety valve. This work could not be completed that day and it was agreed that a temporary seal would be put on the pump until they could continue next day.

When the next shift came on duty they were not made aware of the temporary fix on one of the pumps and when at about 10.00 pm the operating unit failed, they started the back-up pump and high-pressure gas began to leak through the temporary seal. When it finally exploded it blew out firewalls and stored oil started to burn. In twenty minutes the fire had grown and burst the gas risers from the neighbouring platforms. The final massive explosion destroyed the platform, which sank at about 10.25 pm.

Of the 52 who survived all had jumped optimistically off the deck of the burning platform and were picked up by rescue ships. Most suffered from post traumatic stress and did not return to the oil industry. One consequence of this was the creation of the Offshore Industry Liaison Committee which is effectively a trade union monitoring rig safety and maintenance procedures.

In the past thirty-four years other offshore rigs have been lost and people have been killed; but nothing compatible in scale to the *Piper Alpha* disaster has since happened anywhere in the world. The consequence for a few months was a reduction of oil supply from the North Sea and of tax revenue to the British Government. There was no damage on shore and no destruction of any industry.

Nine months after *Piper Alpha*, in the United States, the tanker *Exxon Valdez*, with a cargo of Alaskan crude oil, en route from Valdez to Long Beach, CA, grounded on a reef in Prince William Sound, Alaska, in April 1989 and released 11 million gallons of crude into the sea.

Twenty one years later the sea is starting to recover with first a shrimp fishery, but before the accident a major herring fishery had existed. This has not recovered because the oil spill happened

2.2 Gulf of Mexico: ships attempt to douse the burning Deep Water Horizon rig minutes before it sank in over 1700 metres of seawater, starting a major leak which took four months to drill a relief well to plug it. (Photo courtesy of BP)

at spawning time and the leaking oil smothered the spawning grounds.

So the people in that part of Alaska who had lived by fishing for herring suddenly saw their fishery destroyed. Twenty-three years later a shrimp fishery is reviving but the herring have not returned and their spawning grounds have not recovered.

In April 2010 a much bigger accident saw the US Government pouring blame on to a major foreign oil company. This was due to the explosion following malfunction of the blow-out preventer of BP's *Deep Water Horizon* rig in the Gulf of Mexico. This was a new installation in some 1700 m of water depth, but the explosion destroyed the rig, killed eleven workers, and for the next three months crude oil spewed into the sea until a relief well could be drilled to intercept the damaged well so that it could be permanently plugged.

But this was not so simple. It was an event on the sea bed in an unprecedented depth of water. Various attempts to cap the well and recover some of the leaking oil could only be performed by remote controlled submarines which would be operating under a pressure of about 450 atmospheres. It was not until August 2010 that the damaged well was finally plugged.

In his rage at what had happened President Obama persisted in referring to British Petroleum, although the company had stopped using the name many years ago. In fact the rig had been leased from Trans Ocean and most of the equipment and supplies were provided by Haliburton, both American companies, whose equipment had failed. But BP as the operator, would have to pay for the clean up, the total cost of which, in January 2011 was put at $26 billion.

The Louisiana and Mississippi coast lines were one of the major fishing centres of the country famous for oysters, scallops and other shell fish. Commentators had visions of the fisheries being unusable for decades and a whole industry could be destroyed, so this was another aspect of the compensation to be paid. Fishermen who could not fish were employed by BP to take their boats out to lay kilometres of booms to stop the oil coming ashore and polluting the coastal marshlands.

The death of workers in an industrial accident is a tragedy for their families, friends and co-workers, but generally new people with similar skills can be employed to replace them. But a whole industry destroyed as the result of an accident cannot be so easily replaced, particularly in cold climates where the break-up of oil in the sea is much slower. The condition at the *Deep Water Horizon* site was the extreme depth of water in which the leakage was happening so that not all of the oil came up to the surface, but given that the Gulf of Mexico has a much warmer climate the marine environment may surely recover faster than it has in Alaska.

Rising Arctic temperatures have probably returned Greenland to the climate at the time of the Viking settlements there 1000 years ago. Prospectors have found large deposits of oil and natural gas, uranium, nickel and other minerals. Exploitation of oil and gas in the Arctic is now being talked about as the output of the oil fields further south declines.

Already there are Arctic oil and gas fields in production. On the North Slope of Alaska at Prudhoe Bay are a group of on-shore oil and gas fields discovered in 1968 which had 26 billion bbl of oil and some 740 billion m^3 of natural gas. Production started in 1977 on completion of a 1300 km pipeline down to the all-weather port of Valdez from where waiting tankers shipped it down to west coast ports on the US mainland. The pipeline is entirely above ground mounted on insulated supports so that the temperature of the oil does not melt the permafrost.

Initially the flow rate was 2 million bbl/day but after thirty four years it has declined to 660,000 bbl/day in 2011. There has been one major accident resulting from corrosion of a crude oil pipeline from one of the satellite fields in March 2006. This spilt 5054 bbl over an area of 7000 m^2 and was the largest to date in Alaska. The pipeline was decommissioned and a smaller diameter pipe was installed reflecting the reduced capacity remaining in the field supplying it.

Since corrosion was the cause of the leakage it was decided to inspect the main pipeline down to Valdez. A badly corroded section was found and a 35 km section had to be replaced. Production was halted for two years while the new pipeline section was installed, and production resumed in January 2008.

East of Prudhoe Bay and extending to the Canadian border is an area designated the Arctic National Wildlife Refuge in 1977 on completion of the Valdez pipeline. It covers an area of 510,000 km^2 and is the calving area for a large herd of the porcupine caribou, named after the Porcupine River, which flows between the two countries. For this reason drilling is banned in the Refuge but early seismic surveys indicated the presence of about 10 billion bbl of oil and 26 billion m^3 of natural gas.

Whether or not to drill in the Refuge has ever since been a contentious issue in Washington. Republicans tend to favour drilling. It would be on shore and it might be necessary for national energy security. Democrats, siding with the Inuit and other tribes in the Arctic region, are against drilling and would like the area to remain as pristine wilderness. Deep Water Horizon got people thinking what would happen if a similar event happened in the Arctic.

In the European Arctic three gas fields, Snøhvit, Albatross and Askelad are located 140 km northwest of Hammerfest on the north coast of Norway. Between them they have estimated reserves of 193 billion m^3 of natural gas, 17.9 billion m^3 of condensate and some 5.1 billion t of natural gas liquids.

The shore terminal is on Melkoye Island, near Hammerfest, and comprises a gas separation and liquefaction plant which, because of the extreme northerly location was built in Spain on a barge, and towed up to a dock within which it was permanently moored. This saved a considerable cost and time for construction since there was no issue of major steel erection and shipping all items of plant for assembly on site. Furthermore the wellhead equipment is remote controlled and mounted in pressure containers on the sea bed.

The plant on Melkoye Island started production in 2007 with a single process train fed from the Snøhvit and Albatross fields. Askelad is due to come into production in 2014-15. The first LNG shipments were in October 2007 to southern Europe followed by one to the Cove point, MD, terminal in the United States.

But there have been problems at startup, mainly in the sea water heat exchanger of the cooler which led to a shut down from March to July 2008, and again from June to November for further repairs and upgrades to the cooling system. The heat exchanger was due to be replaced in 2011.

In full production the plant will produce 4.3 million t/year of LNG which will be transported in one of four 145,000 t ships which were ordered with the project. Some 70 shipments/year will serve customers in Southern Europe, and the United States.

One energy industry stands out from all the rest because of its origins: nuclear power. The bombing of Hiroshima and Nagasaki not only brought the Second World War to an end but demonstrated the enormous destructive power of the bombs and the effects of radiation on the human body.

In the United States, the Peaceful Use of Atomic Energy was a central plank of the Eisenhower Administration, from 1952 to 1960. So when during this time the first nuclear reactors were produced for power generation and marine propulsion, this background knowledge inspired the industry to take great care of their workers and avoid excessive exposure to radiation.

Some exposure is inevitable in certain jobs and the maximum exposure rate over time is defined in law. Go do any work in a nuclear station or development laboratory and you will receive a film badge to wear. You may also be required to at least change shoes and possibly some or all of your clothes, depending on whether or not you will be entering a high radiation area.

When you leave the plant your badge is scanned, to see what dose of radiation, if any, you have received. You change back into your own clothes and the overalls that you took off are classified as low-level radioactive waste.

With over sixty years of nuclear technology there are people now alive who have spent their whole working life in the industry, been submitted to regular radiation monitoring, and have retired in better health than the majority of coal-miners. Indeed it is more likely that a former nuclear industry worker would have died of lung cancer because he smoked, rather than because of where he spent his working life.

A nuclear power plant can suffer breakdowns to the turbine and generator just as in any other type of power station with a large steam turbo-generator. If anything like this happened, the control system would automatically shut down the reactor. However it is more likely that a nuclear trip would be due to some external factor on the grid, or an earthquake as has often been the case in Japan. But whenever it happens, the nuclear trip drops all the control rods into the reactor core to shut off the reaction

But if for any reason a nuclear plant trips out, there is a sort of black start capability based on a group of large diesel engines which pick up the auxiliary loads as soon as it happens. The control system is still operating and the circulation pumps are running to remove the residual heat from the fuel in the reactor core. When the external circuits are restored they take over, the diesels shut down, and the reactor can be brought up to full load again.

In more than sixty years there have been only four accidents to operating reactors; Windscale, UK, in October 1957; Three Mile Island near Harrisburgh, PA, in March, 1979; Chernobyl, Ukraine, in April,1986; and Fukushima Daiichi, Japan in March 2011. Of these, Three Mile Island had no serious effect on the surrounding environment, except in the minds of the journalists reporting it.

In Japan on 12 March 2011 the most powerful earthquake since records began (8.9 Richter), occurred 130 km offshore Sendai at a depth of 24 km. The great Kanto earthquake of 1923 was almost 100 times less powerful at 7.3 Richter and killed 140,000 people. Eighty-eight years later it was the following tsunami with a 30 m high wave front that had done the damage, sweeping all before it.

Windscale, later renamed Sellafield, was the site of two air-cooled graphite-moderated natural uranium reactors designed to produce plutonium for the British nuclear weapons programme. The fuel channels were horizontal so that converted fuel could be discharged into a duct leading to the reprocessing plant which would separate the plutonium.

On the morning of 8 October 1957, one reactor core had been annealed against radiation damage and had been fuelled but the annealing temperature had not fallen and the reactor started to heat up. The uranium fuel became red hot and when attempts to contain the fire by trying to push out the burning fuel rods failed and also to pump in carbon dioxide to quench the fire, it was decided to use water. Eventually the fire was doused but not before some gases had escaped carrying radioactive particles into the environment, principally Iodine 131, and Caesium 137.

Iodine is the dangerous isotope, because it can be absorbed by human beings and passed to the thyroid gland to cause cancer, but it has a half life of only 8 days which means that in the space of a month the radioactivity would have declined by a factor of 64. Thus over a 700 km^2 around the site farmers were obliged to milk their cows and then dispose of it.

This arrangement lasted a month; but what was surprising was the public acceptance of what had happened. In fact, a few weeks earlier, HM Queen Elizabeth officially opened the neighbouring Calder Hall nuclear power station, the world's first. If there was any worry about the accident, it was that it might stop the nuclear power program in its track, and even worse, we would have to buy American water-cooled reactors. But the programme continued. Sixteen more reactors were built on seven sites all of the gas-cooled, natural-uranium Magnox design.

Some 21 years later at Three Mile Island near Harrisburg, PA, were two 802 MW pressurised water reactors (PWR) by Babcock & wilcox which had gone into service in April and September 1974. The cooling water pumps on Unit 2 suddenly stopped. The steam turbine could not therefore remove heat from the reactor and automatically shut down. Since this caused the reactor pressure to increase, a relief valve on the top of the pressurizer opened and was meant to close when the pressure had dropped to a defined level.

However the operators received confusing signals. They thought the relief valve had closed when in fact it had stuck open. Second they had no instrument that indicated the water level in the pressure vessel, but since the pressurizer was full they could only assume that the reactor core was completely under water. In fact it was not and had started to overheat and the core had partially melted, but this was not found out until a few months later when it was deemed safe to open the pressure vessel for an inspection.

Windscale had released radioactive gas to the environment; but at Three Mile Island nothing escaped, everything was retained within the containment structure. Nobody at the plant was killed or injured; to the outside world it was almost a non-event. But the American public reaction was totally different.

The Green Organizations were already a force to be reckoned with. Their growing opposition to nuclear power had manifested itself at planning inquiries for new power plants which extended their duration and thus their cost. The utilities could not recover these costs from their customers and in many cases cancelled the planned power plant.

It was unfortunate that it happened when it did because for six years the anti-nuclear fanatics had built up their campaign with the myth of the China Syndrome whereby the melted core would be so heavy and would contain so much energy that it would bore down through the earth and come out the otherside in Beijing.

The Greens had even persuaded their supporters in Hollywood to make a film with the same title China Syndrome, which was released twelve days before the accident. Remember too, that at this time, there was still hostility between the United States and China during the dictatorship of Mao Zhe Dong.

Yet everything was held within the containment building, nobody was killed or injured in the plant or in the surrounding community, and if it proved anything at all it showed that a nuclear power plant (specifically a PWR) could suffer a major nuclear accident and hold everything within the containment.

The plants which were then under construction were completed and put into operation. Unit 1 of Three Mile Island has continued in operation and in 2004 was granted a life extension to 60 years.

2.3 Three Mile Island, PA: site of two 800 MW PWR's of which Unit 2 in the foreground continues to operate 33 years after the accident and has life extension to 60 years. (Photo courtesy of Exelon Nuclear)

It will now be shut down in 2034.

The accident had profound effects on the industry. No more nuclear power plants were ordered until 2009. But more important completion of the fuel cycle was abandoned and construction of a reprocessing plant at Barnwell, NC was halted. Nuclear operators were obliged to run a once-through fuel cycle and store spent fuel on site. They could not send spent fuel to France for reprocessing and accept mixed oxide fuel in return. However mixed oxide fuel reprocessed from retired weapons under the US/Russian weapons reduction treaties, has been burnt in a Duke Energy Power Station.

During the twenty years following the accident it was generally assumed that the United States would never build any nuclear power stations again and shut all the existing units down at the end of their operating lives, assumed to be 40 years. But as the plants aged and their reliability increased it was seen that they had the lowest production costs of any thermal power plant.

The nuclear waste repository at Yucca Mountain, NV has been mired in controversy with a Democratic State Government firmly against having it in their territory. Now the Federal Government of the same political ilk, has decided that waste will not be stored there in what is effectively a completed installation.

Nine years after Three Mile Island, at Chernobyl, Ukraine, in April 1986, one of four Russian-designed RBMK reactors under a low power test was destroyed in an explosion which blew the top off the reactor and its building causing a cloud of radioactive debris to be carried on the prevailing wind up over Byeloruss, and across Northern Europe.

The consequences of this accident were politically profound. First the Green Movements were ecstatic. They had always claimed that nuclear power was unsafe, and didn't this prove it?

Well, no; because the RBMK was a uniquely Russian design which had been studied abroad and been found to have a basic design fault which the unusual test procedure had revealed. But this accident had consequences beyond the Soviet Union because it opened up an unprecedented dialog with the nuclear industries of the United States and Western Europe which collaborated in the inspection of every power reactor, both RBMK and PWR, and the production of control equipment which would bring them up to western safety standards.

There would have been a serious reduction of electricity supply if all the RBMK reactors had been shut down. So it was a priority to find out what had happened and make the necessary modifications to the control and safety systems.

The other big event which Chernobyl undoubtedly accelerated was the collapse of Communist Government in Eastern Europe and the Soviet Union. Indeed that country no longer exists and the only examples of Communist Government remaining are in North Korea and Cuba.

But the interesting thing is that while apart from some cases of thyroid cancer from the short-lived radioactive iodine isotope, mainly among the surviving power plant staff, rescue workers and clean-up operatives, in the twenty-five years since the accident there has been no higher than normal incidence of other common cancers in Kiev and the surrounding districts. Chernobyl is about 130 km north of the city.

The European Union was worried about two particular stations. Chernobyl of course, with three operating reactors remaining; and

Ignalina, Lithuania, where there are two 1500 MW RBMK. When, after the break up of the Soviet Union in 1992, the three Baltic republics (Estonia, Latvia, and Lithuania) filed applications to join the European Union, one of the Ignalina units had been shut down in 2004, but the other continued in operation until December 2009. It is planned to replace them with two western reactor types of the latest design.

No more RBMK reactors have been built and the remaining plants in Russia are still running. However, new nuclear plants in Russia will be of a new design of 1000 MW PWR. Among the first to be installed are two at Sosnovy Bor, near St. Petersburg which will enter service in 2012 and 2013. This is the first step in the replacement of the four RBMK units on the site.

A Japanese repeat of Three Mile Island at first seemed likely in the aftermath of the 2011 Sendai earthquake. Four nuclear plants on the east coast of Honshu, Higashidori, Fukushima Daini, Onagawa and Fukushima Daiichi automatically tripped out as intended, but the southernmost plant: Fukushima Daiichi was severely damaged by the following tsunami. The stations all have Boiling Water Reactors which are directly coupled to the steam turbines. The new reprocessing plant at Rokkasho Mura, was also tripped out and successfully switched to its back-up diesel generators.

The biggest problem was at Fukushima Daiichi. Three of the six reactors were shut down for maintenance, and their fuel had been moved to fuel storage ponds. The three operating reactors, although they shut down automatically as intended the earthquake had cut off the grid supply leaving them to depend on the stand-by diesels. But these failed after an hour as the tsunami reached the coast and flooded them so that use of the cooling pumps was lost. Back-up batteries came into action, but pressure was building up in the reactors and had to be relieved by blowing off steam.

Some 24 hours after the earthquake there was an explosion in the reactor building of unit 1 which demolished the roof and walls but left the structural frame of the building intact and the reactor in its containment vessel. Two days later a further explosion followed at Unit 3 which was put down to hydrogen from reactions when blowing off to reduce reactor pressure. The question remaining was whether the reactor cores of the three units were uncovered.

It was decided to flood the reactors with sea water to keep them cool.

It was then found that the spent fuels storage pond of Unit 4 was leaking and if allowed to empty the spent fuel could overheat and release radiation. Because of the design of the BWR steam discharge would have released some radiation, principally Iodine131 and Caesium137.

In the subsequent discussions on radio and television around the world nothing was said about the three other nuclear plants that had shut down normally. Further up the coast Fukushima Daiini had successfully brought its four reactors down to cold shut down in three days. Since the tsunami took out the transmission lines the plant cannot start up again until they have been replaced.

Of course March 2011 saw all the anti-nuclear Green fanatics coming out of the woodwork and appearing on international radio and television networks where they were introduced as nuclear experts. In broadcasting nothing seems to have changed in thirty years! But did these people know what they were talking about? When a Green fanatic tells Al Jazeera television that plutonium is only suitable for nuclear weapons one seriously wonders how much these people understand of nuclear energy: not least that Fukushima Daiichi Unit 3 was the first Japanese reactor to use mixed oxide fuel.

When a nuclear plant trips out the control rods all drop into the reactor to stop the reaction. The Unit 1 explosion did not happen until 24 hours after the earthquake, so for all that time there had been no reaction taking place. Had it been a nuclear explosion would the steel structure of the reactor hall still be standing, and would there have still been a television camera standing to show us what happened after the explosion?

Furthermore it was clear that effective cooling of the reactor cores and the spent fuel ponds at Fukushima Daiichi could not be achieved until a power line could be re-erected to bring a back-up supply to the station. The problem was complicated by earthquake damage to the low voltage circuit panels. However by Saturday March 19, a cable had been connected from the nearest working substation to four reactors so that cooling of the undamaged spent fuel ponds could be started again. The fire brigade had been spraying water into the damaged pond for two days.

All the BWR's shut down were early model reactors and all installed before 1995 with two or three external pipe loops to the large circulating pumps. It was not until 1995 that the first ABWR was installed. This advanced design has ten small circulating pumps in the bottom of the reactor vessel with the nacelles of the pump motors hanging down as an integral part of the structure. Two of these reactors were installed at Kashiwazaki Kariwa and were found to have withstood greater ground accelerations than those for which they had been designed, after the 2006 earthquake which had its epicentre only a few kilometres away.

But it was enough to get the Europeans worried and Germany ordered the closure of seven reactors installed contemporary with Fukushima Daiichi for a safety inspection. Given the political complexity of the future of nuclear power in Germany, that was probably all that Prime Minister Angela Merkel could do to protect her pro-nuclear position. There are neither historical records of earthquakes in Germany nor of tsunami on the Japanese scale in the Baltic. In any case no German nuclear power plants are on the coast, although Brokdorf is on the Elbe estuary a few kilometres downstream from Hamburg.

Several countries are looking at nuclear safety, particularly in regard to older reactors. It is easy to forget that, although the first expansion of nuclear energy came in the 1970s and 80s, most of the existing plants have been more than thirty years in operation. But the Japanese earthquake and tsunami were of unprecedented severity which only a few countries are ever likely to experience even at coastal plant sites.

Perhaps the most dangerous energy source is not uranium but coal, specifically deep mined coal, much of which also contains methane and a certain amount of radioactivity which comes out in the flue gases, and which nobody worries about. In the early days of mining in Britain, miners would hang a cage containing a canary in the mine. If the bird suddenly stopped singing and died it meant that there was too much gas present. More reliable instruments now monitor the presence of gas in mines.

The worst accident in Britain occurred on 14 October 1913 due to an explosion in the Universal pit at Senghenydd, in South Wales when 439 miners were killed. This accident is a record for the number of miners, killed in one event. The records of other mine accidents show that usually fewer than 100 people are killed and more often less than half that number.

But this pales into insignificance against the Chinese record where the mines are considered to be some of the most dangerous in the world, with each year a large number of accidents each with a relatively small number of casualties. In 2009 there were 1616 accidents in Chinese coal mines which killed 2631, which was 18% fewer than in 2008.

Most mine accidents are due to explosion of entrained methane and the consequential rock falls. The majority of casualties are those trapped by the rock fall or actually killed by it. But with the closure of many deep mines in the years after the 1984-5 strike, there has been in the UK a resurgence in opencast mining, which now accounts for the majority of production.

Deep mined coal continues in many parts of the world but a few schemes have started to extract methane which is used on the surface for heating water for pit-head baths, and buildings on the surface. To extract methane before mining can begin has obvious benefits in mine safety.

In central Queensland there are large deposits of coal which are too deep to mine, but it has been possible to drill into the seams and bring the gas up to the surface where it is compressed and fed into the public natural gas system. Consumers in and around Brisbane burn a mixture of natural gas and coal mine methane.

Coal must have a future as a chemical feedstock, but in the mean time it is in decline. The Large Combustion Plant Directive will cause a large number of European coal-fired power stations to be shut down at the end of 2015.

Then as the nuclear revival develops it is not the addition to and replacement of existing nuclear plants in Europe and North America so much as with a group of countries in the Middle East and Asia that have not had nuclear power stations up to now. These are countries with very little or no coal-fired capacity, and high rates of growth of electricity demand: Turkey, Egypt, United Arab Emirates, Indonesia, Vietnam and Thailand are all planning their first nuclear power plants. They see nuclear power plants as way of adding to their electricity supply without increasing their greenhouse gas emissions.

So if reducing emissions is to guide our choice of energy, it will obviously be nuclear, a technology which Green activists believe to be inherently unsafe, but which has seen in its sixty year history, only three significant accidents, two of which produced no fatalities. And one earthquake followed by a tsunami which prevented the cooling down of three reactors on the Japanese coast.

But without the investment in nuclear power how do we supply the energy demand which is coming as we electrify more of our transport system and move to hybrid, or all-electric cars, and at the same time cut our greenhouse gas emissions?

We will still be moving crude oil around for many years. It is no longer a primary fuel for power generation, but it will take a much longer time to make a significant reduction in widespread use for marine, air, rail and road transport.

If the price of oil continues to rise, what effect will that have on the general public as to their use of cars and the future strength of the market? What effect will it have on the speed with which new power plants are built?

3
New role for combined cycles

For the last twenty years the gas-fired combined cycle has been the preferred power plant for capacity additions over much of the world. It has low environmental impact, high efficiency, and is quick to build. Furthermore, since the primary power unit is a gas turbine, it has greater flexibility of operation.

Inevitably we will still need some fossil-fired power plants because of the daily variations in electricity demand. Nuclear plants have poor load-following capability; there is not enough water power potential remaining in the developed industrial countries; and the new renewables, solar, marine currents, and wind, cannot operate continuously and therefore cannot be always available to meet peak demand.

The beginning of combined cycle development started about ten years after the end of the Second World War. Gas turbines began as aero engines burning kerosene, a light distillate fuel. This was followed by the development of more robust gas turbines which could not fly, and which actually burned oil or natural gas, to drive pumps and compressors on oil and gas pipelines.

These early gas turbines were small, high-speed machines and although a few were used for power generation applications it was unlikely to continue unless larger gas turbines running at the common synchronous speeds could be developed.

During the 1950s electricity demand was growing rapidly as nations rebuilt and developed new industries after the war and population grew. There was great concern over the efficiency of power generation which was around 30% and generated entirely by coal- or oil-fired steam plants, or by water power in those countries that had it.

In the southern United States some schemes were built in which gas turbines were used as feed water heaters for steam plants. This raised the efficiency to over 40%. The gas turbine had a heat recovery boiler which replaced some of the feed water heaters. Normally these feed water heaters are connected to steam bleeds from the turbine, thereby reducing the amount of steam available to make electricity. Some of these heaters were replaced by the gas turbine heat recovery boiler, thereby increasing the output of the steam turbine. But only six plants of this type were built.

In Europe the 75 MW power station, completed in 1960 at Korneuburg, Austria, was of a different design with two 25 MW gas turbines each with its own heat recovery boiler feeding into a common range supplying a steam turbine. Because of the low exhaust temperature of the gas turbines, burners had to be installed in their exhaust ducts to raise the steam temperature leaving the boilers. Since these gas turbines did not run at the synchronous speed they had to drive their generators through a reduction gearbox. But this was the most efficient thermal power plant at that time and it ran for fourteen years as a base-load station.

Undoubtedly the most important development to advance the combined cycle technology was the production of large gas turbines which could run at synchronous speed and therefore eliminate the gearbox to the generator. In 1968 the first large gas turbines to run at 3600 rev/min had appeared in the United States. Five more running at the European synchronous speed of 3000 rev/min were available by 1975. It is from this time that combined cycle development really took off.

While this was happening, war broke out between Israel and its Arab neighbours in the autumn of 1973 which saw the mainly Arab, Organization of Petroleum Exporting Countries (OPEC) impose a 4-fold increase on the price of a barrel of oil. With, in Europe and North America, so much oil being used for power generation and a growing car market this was a crisis.

The decision was taken in the developed industrial countries to take oil out of power generation. The alternatives were coal, nuclear energy and natural gas, which had already been discovered in the North Sea and on the North Slope of Alaska. In the following years many new gas fields were discovered around the world and this immediately benefitted the developing countries of south east Asia, who in the 1980s were the first large market for combined cycles, particularly in Japan, Thailand and Indonesia.

The first of these larger gas turbines to be used for combined cycles, and collectively known as the E-class, were Brown Boveri's Type 13, GE's Frame 9E, Mitsubishi's MW701D, and Siemens' Model V94.2 in the 50 Hz market. Outputs ranged from 90 to 150 MW and with exhaust temperatures over 500°C which could provide a high pressure steam cycle without the need for an afterburner. The basic combined cycle was in 2+2+1 arrangement, i.e. two gas turbines powering one steam turbine which gave a total rating of 360 to 450 MW at an efficiency of about 43%.

The next development was to split the steam path through the boiler. The high pressure section was heated by exhaust gas entering the boiler and the low pressure section took the remaining heat available up to the base of the stack. In this two pressure arrangement high pressure would be about 75 bars, 510°C and the low pressure about 6 bars, 200°C; this for 100 MW gas turbine. All three generators would be the same size. Best efficiency obtainable with these E-class gas turbines was about 52.5%.

The disadvantage of the 2+2+1 arrangement is in the much lower efficiency in load following when, for example, at night one gas turbine is shut down. The steam turbine only has half the steam supply and is therefore running at part load. To improve this, one or two plants have been built with three and even four gas turbines powering one steam turbine. In Indonesia the standard format for combined cycles with E-class gas turbines was a 3+3+1 arrangement.

In Thailand, Khanom, in the south of the peninsula near the Malaysian border has four GE Frame 9E gas turbines and a single 220 MW steam turbine. There are few power plants in the area and the arrangement allows load following by shutting down one or two gas turbines at night. The alternative to this would have been high-voltage transmission lines more than 1000 km long, and with corresponding reactive power losses, coming down the peninsula from Bangkok.

Saudi Arabia has also built combined cycles in 4+4+1 arrangement with E-class gas turbines, notably at Rabigh, north of Jeddah on the Red Sea coast, which entailed the conversion of four simple cycle gas turbines into two combined cycle blocks; and Riyadh Power Plant 9, where there are four 350 MW blocks each with four GE Frame 7EA gas turbines. All gas turbines in Saudi Arabia burn treated crude oil and are designed for operation in temperatures over 30°C. In choosing the 4+4+1 arrangement they got higher efficiency and flexibility for base load power.

Limay Bataan was the first combined cycle built in the Phillipines. Completed in 1995 it has two blocks in 3+3+1 and is oil fired. Situated about 100 km northwest of Manila, it is used primarily as a peaking plant. During construction the Malampaya gas field was discovered offshore southwest Luzon. The gas was to come ashore at the Shell Refinery about 20 km south of Batangas. Three more combined cycles were built in the south and they are all gas fired. The first of these, Santa Rita, is a 1000 MW plant by Siemens with four 250 MW single-shaft blocks with their Model V84.2 gas turbines. The plant went into operation in 2002, but for the first two years it had to run on liquid fuels until natural gas became available. On an adjacent site the 500 MW San Lorenzo plant was also built by Siemens with two 250 MW blocks and came into operation in 2004 after gas had arrived on site.

The third combined cycle was built at Ilijan, beside a deep water anchorage, on the south coast. Accessible from Batangas along a 100 km, poorly maintained jungle road around a mountain, the site literally had to be carved out of the jungle. There was none of the basic infrastructure on site, which all had to be installed before anything could be built. First, a dock to receive everything else was required on the site. Water supply from came from reverse osmosis units but the gas supply was brought over the mountain from Batangas.

The combined cycle by Mitsubishi was the first application of their steam-cooled M501G. There are four gas turbines and two steam turbines forming two 700 MW three-shaft blocks. The plant was built under a Build Operate Transfer (BOT) system which runs for 20 years, ending in 2022. Korea Electric Power Corporation supervised the construction and now operate and maintain the plant; they also train local people as operators. Power is sent out over the mountain to Batangas and on to Manila. Limay Bataan still runs on oil, but the other combined cycles are now gas-fired and hold an oil supply only as a back-up fuel in the event of a problem with gas supply.

Until 1990 the combined cycle market was mainly in Japan and south east Asia. Then as Europe, and the rest of the world, started deregulation of electricity supply which separated generation from transmission and distribution, it opened the market to private generating companies who greatly expanded it. In the United Kingdom a large number of combined cycles were built to replace old, small, coal-fired plants so that today 40% of the electricity supply is from gas-fired combined cycles.

The subsequent increase in combined cycle efficiency to almost 60% can be put down to higher steam conditions and greater gas turbine efficiency. Just before deregulation the first F-class gas turbines were announced in 1989. These were rated 180 MW in the 60 Hz market and 220 MW for 50 Hz. But with exhaust conditions approaching 600 kg/s at 600°C there was enough energy for a tri-pressure reheat steam cycle, for which the efficiency was quoted at 55%.

One concern of higher temperatures was whether or not the low nitrogen oxide (NOx) levels achieved with the E-class gas turbines would be maintained. The other issue was how the gas turbine fabric would stand up to higher firing temperatures which are now close to 1300°C on the largest machines.

During the 1980s the gas turbine industry had concentrated on the development of low emission combustors. The aim was to remove NOx formed from the oxidization of the excess combustion air at firing temperatures above about 850°C. Much lower NOx emissions down to less than 10 vppm had been reported on some E-class machines. But more significantly, by the end of the decade, a contract principle was established whereby all gas turbine plants would guarantee NOx emissions on dry gas of 25 vppm, and on oil at 42 vppm with water or steam injection. In practice NOx levels less than15 vppm have been achieved on the largest gas turbines.

The basic environmental measures had been established for the combined cycle with the gas turbines of the time, and with the reduced competitiveness of coal due mainly to the compulsory fitting of FGD to all new power plants, the combined cycle at this

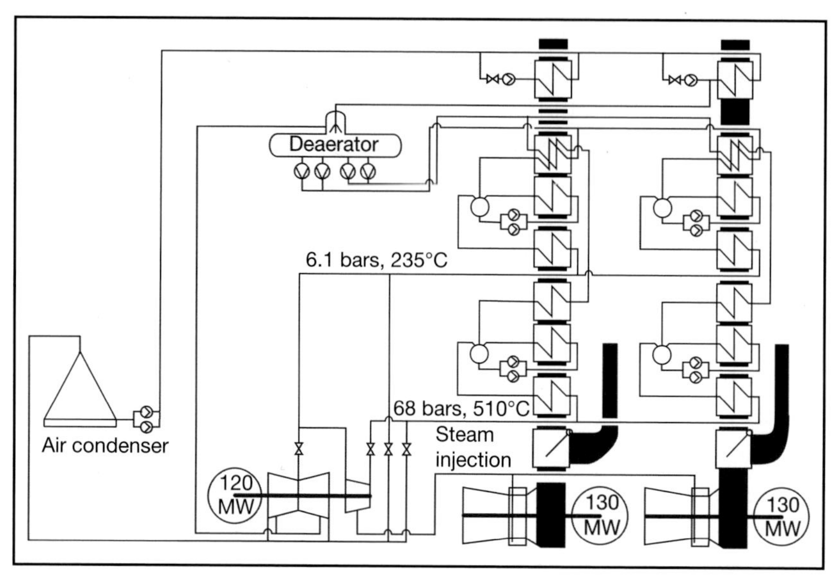

3.1 Peterborough, UK: schematic of an early model combined cycle one of several plants owned by Centrica, with GE Frame 9E gas turbines and dual pressure steam cycle with an efficiency of 48%.

time became the preferred equipment for capacity addition. The contract condition for NOx emissions continues to the present day, and only one gas turbine, Siemens' 45 MW model SGT800, can meet the NOx condition for oil without water or steam injection.

As the market developed during the 1990s, it was the F-class gas turbines from the European manufacturers which came to the fore. Whereas both American companies had adopted can annular combustion systems with small combustion cans arranged in a ring around the shaft with the gas flow splayed out in transition ducts leading to the power turbine. The first Brown Boveri design had a single combustion chamber mounted on top of the casing and containing a single large burner. Siemens design had two vertical combustion chambers, one either side of the machine, with a ring of six small burners in the top of each casing.

These E-class gas turbines therefore had a large steel mass in the centre which directed the compressed air up to the combustion chambers, and the returning hot air into the turbine inlet. As temperatures rose their solution was to use an air-cooled annular combustion chamber similar to that of the large turbofan engines which by then had been in airline service for over twenty years.

3.2 Limay Bataan, Philippines: two 300 MW oil-fired blocks are the first combined cycle installed in the country, north of Manila on the main island of Luzon in 1995. (Photo courtesy of Alstom Power)

At the same time both manufacturers simultaneously introduced scaled designs for both system frequencies. Siemens have since retained only the 50 Hz model in production and use an equivalent Westinghouse design as their current 60 Hz model.

In 1993 ASEA Brown Boveri (ABB) brought out upgrades of their E-class models which were followed in 1994 by their two F-class designs, GT24 initially at 160 MW for 60 Hz and the GT26 at 240 MW for 50 Hz networks. The particular features of these machines are a high pressure ratio of 32:1 and sequential combustion.

In this arrangement there are two annular combustion chambers separated by a single turbine stage. In 1997, as the prototype GT24 went into service at Jersey Central's Gilbert peaking plant, a GT26 test started at the Birr works in Switzerland to study the performance envelope; subsequently, GT24 was uprated to 183 MW and GT26 to 265 MW.

Clearly with these unique designs ABB were looking at future applications. With sequential combustion the fuel flows to the two combustors can be controlled individually, so that it is possible to hold the exhaust temperature at the full 640°C down to about 40%

of gas turbine load while only reducing the fuel to the primary combustor. As a result the efficiency at part load is higher, with better economy for load following.

The introduction of these gas turbines did not go smoothly. First, on GE's Frame 9FA, excessive rotor vibration was observed after 1000 operating hours. Not only that, but in the 60 Hz market because of the higher speed and smaller dimensions of the Frame 7FA the same problem did not appear until about 10 000 operating hours.

At the time, the Anglo-French company GEC Alsthom was still a gas turbine licensee of GE and at Belfort, in Alsace, were the production lines for the large gas turbines of European Gas Turbines (EGT) a jointly-owned company. The small gas turbines for industrial combined heat and power, and pipeline duty were made in the UK at their Lincoln factory. With so many units to be repaired GE and EGT devised a repair procedure and then licensed it to GE service bases around the world.

The large number of gas turbines involved had arisen because customers had been beguiled by the prospect of 55% efficiency which would significantly cut their fuel costs and had quickly placed orders. Besides the operating gas turbines there were those under construction. For example, at EGAT's South Bangkok station the erection of the bottom half of the casing could continue for both units until the modified rotors could be shipped up from GE's Singapore maintenance base.

ABB also came to grief with the upgrades of their GT24 and GT26 models. This started as a problem with blade cooling of the high pressure turbine stages, but in a way it was a less serious problem, being thermal rather than mechanical, than that which had struck GE. Many ABB customers had installed either GT24 or GT26 on the understanding that an upgrade was in prospect and had designed their plants accordingly. The heat recovery boilers were designed for the higher exhaust gas flow and similarly the steam turbine was optimised to the projected final rating.

The sequential combustion design had attracted many merchant plant customers in the United States and while this problem only arose with the uprated engine, it did not require a strip down and shipment of the rotor back to Switzerland to fix it. With variable inlet guide vanes, all their customers had to do was to partially close the inlet so that the turbine ran effectively at part load. They could continue like this until the new blades and other design improvements could be installed at a scheduled maintenance outage.

The F-class gas turbine problems led to significant changes in the industry. GE sought to consolidate their business in Europe and took over their Italian business associate, Nuovo Pignone which became GE Oil and Gas. Alsthom were not willing as a major French engineering company to be taken over by GE and gave up their licences. However GE retain the production line at Belfort and their design offices which are in a completely separate site on the other side of a main railway line.

Having separated from GE, Alsthom dropped the letter "h" out of their name, becoming Alstom, as it has always been pronounced. Now without a large gas turbine business they joined up with ABB as ABB Alstom, which subsequently broke up with Alstom retaining the power generation and transport businesses, and ABB becoming an electronics company, focusing on power plant control systems and industrial automation.

Alstom had taken on the GT24 and GT26 problems and had run up an enormous financial loss in 2002-3 not so much because the gas turbines did not work, but rather in the legal settlement of compensation claims for units falling below guaranteed output and availability, and the setting up of a concentrated development program to refine the design and bring it up to then current B rating. This gave rise to rumours that Alstom would soon be closing down. So they started a program of asset sales which included both the Lincoln factory and another one in Finspong, Sweden; making small gas turbines for the industrial market. These were both sold to Siemens who were then seeking to enlarge their gas turbine portfolio, as until then they had no small industrial units in production.

As the problems of the big machines were solved it raised certain questions as to how the turbomachinery industry operated. Traditionally when a new gas turbine was developed the company was only able to perform no-load tests in the factory, and would therefore look for a customer in their own country, if possible,

3.3 Taranaki, New Zealand: steam turbine of the first single-shaft combined cycle with GT26 installed by Alstom and completed in 1998. The turbine has a straight through exhaust to the condenser in the background.

to whom they could sell the prototype unit on the understanding that they would have access for the first six months to supervise operation and make such adjustments as were necessary to ensure that the plant would continue to operate as intended.

Of the major suppliers only Siemens could test the gas turbine on load in the factory. In their Berlin assembly plant they had installed a water-brake which was progressively upgraded so that they could test all their gas turbine models on load. Siemens' final F-class gas turbine range has two models: the 3000 rev/min SGT5 4000F currently rated at 292 MW; and the 3600 rev/min SGT6 5000F rated at 208 MW.

So there are now effectively four basic designs of gas turbine for combined cycles which stem from the past arrangements of design licences. These are shown in the table. GE had a group of Business Associates to whom they supplied noble parts which the Associates put into their own casings. They also had responsibility for the driven units. Westinghouse had two licensees of their designs: Fiat TTG in Europe and Mitsubishi Heavy Industries in Japan, and development was shared between the three companies.

When in 1996 Siemens bought the non-nuclear energy interests

3.4 Otahuhu, New Zealand: this plant near Auckland has a 410 MW single shaft block seen here looking towards the steam turbine with the condenser in the background. (Photo courtesy of Siemens)

of Westinghouse, an arrangement was made with Mitsubishi to continue production of the Westinghouse designs which they had helped to develop and in which they therefore had certain intellectual property rights. Fiat dropped out of the arrangement, and no longer manufacture heavy-frame gas turbines.

With the take-over completed, Siemens sought to ensure that two vastly different gas turbine technologies were fully understood by production staff in both countries. It started with the construction of examples of the Westinghouse W501F in Berlin and later incorporating some of Siemens design ideas into the American machine. They also produced a design for a single-shaft W501F combined cycle incorporating a Benson boiler. Finally a common nomenclature was adopted for the gas turbine models of both companies: SGT5-4000F is the former V94.3 for 50 Hz, and SGT6-5000F was the W501F for 60 Hz. Now with the production of the H-class gas turbines, the 50 Hz model is being manufactured in Germany and the 60 Hz model at the former Westinghouse factory in Orlando, FL.

The F-class gas turbines in their current ratings are still the most widely used for combined cycle duty. They have almost reached

60% efficiency with values between 57.5 and 59% quite common. The big development that brought this about was the single-shaft block, first introduced by the European manufacturers, with the gas turbine and steam turbine driving at opposite ends of a large generator. Furthermore two or three of these combined cycles can be installed on one site. With this, there would be a common control room, a shared fuel supply system and a common spare parts inventory.

In fact the combined cycle is now almost a standard item with the same gas turbine, generator and steam turbine at a number of different sites. Spanish utilities, for example, have ordered the same single-shaft power train for projects in their own country as well as for plants in Central and South America where they have commercial ties with several utilities in their former colonial empire. The only difference from site to site would be in the heat recovery boiler, whether drum type or Benson, horizontal or vertical; and the cooling system, whether direct by a river or lake, or with wet, mechanical-draught, or hybrid tower assemblies, or an air condenser for each block.

In the single-shaft combined cycle the generator is driven from the compressor end of the gas turbine. There is thus a straight path through the exhaust duct from the turbine to the heat recovery boiler. At the other end of the generator there is first an SSS clutch which separates the steam turbine at start-up. The steam turbine can have either a straight through low-pressure stage exhausting to an end-mounted condenser, or a double-flow stage with a side-mounted condenser.

The clutch is particularly important because the steam turbine is held at rest while the gas turbine accelerates to synchronous speed and is held there while the boiler heats up. There is no need for a separate boiler for gland sealing steam because this can be taken from the heat recovery boiler when it has the right temperature and quality. The steam turbine then accelerates on the full steam output of the heat recovery boiler. As soon as the speed reaches synchronous, the clutch closes to complete the single shaft and loading continues.

But the clutch has another very important function which is all the more so because of the increasing demand for high availability, which underpins the merchant plant trade. If the plant shuts down, for whatever reason, but particularly for repair or maintenance, the steam turbine will take up to three days to cool down and come to rest on its turning gear. If it is rigidly coupled to the gas turbine and generator, then neither machine can be accessed for maintenance until the steam turbine has come to rest. With the clutch gas turbine and generator can be shut down in less than two days.

In any combined cycle the gas turbine is the most maintenance intensive element of the power train. Annual maintenance entails a detailed hot gas path inspection of blades and combustors a lot of which can be done with borescopes, but only with the gas turbine at rest and, where necessary, replacement and repair of various components. Therefore, with the steam turbine isolated on its turning gear behind the open clutch at least one to two days work can be accomplished on the gas turbine and generator before the steam turbine has come to rest.

That is the power train of the single-shaft combined cycle which is effectively a standard item. But the heat recovery boiler can be chosen by the customer, which may be from a supplier in his own country or from an affiliated company of the turbomachinery supplier. Whether to have a cooling tower or an air-condenser will depend on the topography of the site. If it is sited beside a big river or lake, direct cooling from the water body is a high efficiency option.

Single shaft combined cycles have the advantage that they can be installed where the demand is and not all in one place. For example a 1200 MW coal fired power station has reached the end of its life and is to be demolished. Because of the availability of gas and the fact that emissions are lower four 400 MW single shaft blocks will be built. Two will be installed on the site of the old station, and one each will be placed elsewhere on the network.

All four are identical and thus share a pool of spares. Moving around the system there may be no major transmission work to connect them. Furthermore the much higher efficiency means lower fuel costs, and lower emission costs if they come to be imposed at some time in the future.

But the big advantage of single-shaft is its potential for load following at higher efficiency. First, the generator on an F-class

3.5 Gent Ringvaart, Belgium: 360 MW single-shaft combined cycle in a country with a large nuclear base load. Runs weekdays with reduced power output at night and shuts down at weekends.

system is about 400 MW. A 400 MW combined cycle on a 2+2+1 configuration would have three 135 MW generators driven by two less efficient gas turbines and the same sized steam turbine.

An interesting example is at the Gent Ringvaart, station of Société de Production d'Electricité in western Belgium. This is a single-shaft combined cycle with an early model GE Frame 9FA gas turbine and a total installed capacity of 360 MW. It entered service in 1997 and has operated since on a load following regime in a country which derives 60% of its electricity supply from seven nuclear reactors.

Generally the plant operates from Monday morning to Friday evening and for fifteen hours a day from 7.00 am to 10.00 pm runs at full load which includes 25 MW of spinning reserve for frequency control. A typical daytime base load is 350 MW when the efficiency is 55.7%. For nine hours overnight the output is cut back to 200 MW and the efficiency drops to 52%, which is virtually the base load efficiency of the previous generation of combined cycles in 2+2+1 arrangement operating continuously.

One particular feature at Gent Ringvaart is the air condenser. On hot summer days the output may be restricted by the steam turbine,

3.6 San Roque, Spain is fitted with fogging system which gives power boost on hot summer days, and reduces compressor inlet temperature from a high ambient of 40° to 21°C. (Photo courtesy of ENDESA)

so SPE have installed sparge pipes under the condenser which spray fine water droplets into the airflow to reduce its temperature and so avoid a restriction of output due to insufficient cooling.

This shows how it is possible to fine tune the output according to the time of year and the immediate demand for electricity. This is because the working fluid is air which is varying in temperature from day to day. In the depths of winter when the daytime temperature is less than 5°C and there is frost at night, the air is denser and therefore the mass flow through the gas turbine and the output are greater. In the summer when the maximum temperature may be over 30°C the air is less dense and the output of the gas turbine is lower.

The other way of fine tuning the performance of the combined cycle is by fogging which increases the humidity and lowers the temperature of the air flow into the compressor. Fogging was pioneered by Mee Industries in the United States of which a large area of the southwest is desert with low humidity.

The basic fogging system is a series of nozzles mounted behind the air intake filters. Water is pumped to the nozzles at high pressure so that they break up the flow into millions of microscopic droplets. Collectively these have a high surface area to react with the warm air and evaporate. The latent heat of evaporation is given up by the air which cools it.

An example of this at work is ENDESA's San Roque site about 20 km northwest of Gibraltar, in Southern Spain. San Roque is an unusual site in that one block is owned and operated by ENDESA and the other by Gas Natural. Effectively it is two separate power stations; a similar arrangement between the two companies is at San Adrian de Besos near Barcelona.

The intake fogging system was supplied by the Swiss company AXEnergy. The object is to raise the humidity of the air entering the gas turbine which reduces the temperature. On a day when the ambient temperature is 40°C the fogging can reduce the effective compressor inlet temperature to 21°C the mass flow is increased to give about a 5% gain of power. The fogging system is only run when the plant is at maximum output, as on a hot day with low relative humidity and the untreated output of the plant is at its lowest.

The fogging system is basically a power augmenter and is separate from the compressor washing system which uses a much coarser water droplet laced with detergent to clean the inlet guide vanes and blades.

Internal cleanliness can only be as good as the intake filters. Depending on where the combined cycle is sited, fine particulate matter can pass through the filters and settle on the compressor blades and vanes. A plant sited on an industrial estate, or besides a busy motorway may pick up particles from traffic exhaust or from process emissions. On a rural site at certain times of the year there is fine dust and pollen, and on a coastal site there is wind blown salt spray.

All these pollutants can lead to a rapid loss of power output unless the compressor is frequently washed. This can be done off-line or on-line. On-line compressor washing has been used for the past thirty years. The basic system consists of a ring of nozzles set close to the gas turbine inlet flange. A preparation unit mixes water with a detergent and pumps it to the intake nozzles. It is a low pressure system running at about 4 bars, which produces droplets in the range of 50 to 250 microns.

The air-assisted nozzle shown in Figure 3.9 has a flat water spray which is sandwiched between two high-velocity flat air sprays. These protect the water spray and punch it through the boundary layer thereby preventing premature deflection and achieving a longer penetration trajectory. For the largest gas turbines the air-assisted nozzle delivers a controlled and stable water spray pattern in the air inlet stream with more uniform wetting of blade surfaces.

Gas turbine operators as a matter of principle, perform a compressor wash at every shut down. But for a merchant plant which depends on high availability with possibly daily start and stop, on-line washing is essential. Suppose that without on-line washing the gas turbine output falls by 100kW per day which must be corrected when 3 MW has been lost. This would mean twelve outages per year, probably on a weekend for off-line washes.

With on-line washing the compressor will be rendered about 80% clean, and therefore at 20 kW/day of lost output it would take five months before the unit would have to be shut down for an off-line wash. So while there is some loss of output, on-line washing does not require the gas turbine to be shut down so often and therefore its availability is higher.

F-class gas turbines still account for the majority in production. Combined cycles currently under construction in Europe may have to be run on base load until at least 2020, but thereafter used for load following and frequency control. The European Union's Large Combustion Plant Directive (LCPD) requires any steam plant built before 1987 to be shut down unless it has been fitted with flue gas desulfurization (FGD) and other emission control measures. A plant due to be shut down in 2015 will be at least 28 years old, and has already been limited in its operating hours. Also several countries may by then have started to build new nuclear plants which will be in service between 2018 and 2020.

A common variable in a combined cycle design is the method of cooling. In the United States, particularly, the use of air condensers removes the issue of water abstraction for cooling even outside of dry areas of the country. An air-condenser typically is two radiator-type coolers forming a triangle above a base carrying the cooling fan. An installation for an F-class gas turbine would have twenty-

3.7 Peterborough power station, UK: the on-line compressor washing skid with, inset, an air-assisted nozzle in the gas turbine intake. (Photos courtesy of Turbotect Ltd)

five modules in five parallel rows fed from a common header.

Combined cycles with air condensers have the lowest net efficiency because of the large auxiliary load of the cooling fans. However since they are blowing ambient air it is possible to vary the number of fans in use. In high summer all would be operating, but in winter when the air temperature might be between -3 and +5°C, it may only be necessary to run sixteen fans to achieve the same cooling load. Therefore the auxiliary load is lower and more electricity can be sold at a time of peak demand.

Wet mechanical draught cooling towers are giving way to hybrid systems since they reduce the visible plumes which might give rise to patches of fog in critical places. In northwest England the first installations were at the Deeside and Connagh's Quay power stations which were on opposite sides of the Dee estuary and near to a new high road bridge over the river.

In the UK at least there are very few days when cooling plumes are visible. The hybrid system looks very much like the wet mechanical draught towers, but is slightly taller. The condensate enters at the top and descends through a dry section cooled by the air being drawn up by the fan. At the bottom of the dry section the

3.8 Samurinda, Indonesia: at this 66 MW combined cycle about 20 km north of the town in East Kalimantan the gas turbines are two Rolls-Royce RB 211 aero derivatives. (Photo courtesy of Rolls-Royce)

hot water enters a conventional wet section which further reduces the temperature returning to the boiler. The reduced plume is the result of moist but cooler air rising through the hot dry section.

While the early problems of the F-class gas turbines slowed down the combined cycle market there remained a buoyant market for the large aero-derivative gas turbines, both as compressor drivers on pipelines and for industrial combined heat and power schemes, particularly in the United States.

Among the manufacturers of the heavy frame utility gas turbines, only General Electric had an aero engine production line and already had supplied some of their aero-derivatives to industrial power schemes. Meanwhile the other two aero engine companies had for a long time been in technical cooperation with the other manufacturers, who had offered a few aero engines for generating sets in their traditional markets.

It is the large turbofan engines which are now used, specifically the core engine without the fan, which offer generator drives from 28 to 50 MW. GE's LM 6000 has a speed of 3600 rev/min and an output of 42 MW, which is a cheaper installation in 60 Hz markets because it does not require a gearbox to drive the generator.

The largest Rolls-Royce aero-derivatives are the RB211-GT62 at 4800 rev/min and 29.89 MW. A new enlarged version introduced in 2012 is the RB211-H63 which like the larger Trent engine has dry (DLE) and wet (WLE) versions, the latter with water injection. The RB211 H63 DLE prototype went on test in Montreal, Canada at the end of 2011. The gas turbine is rated 44 MW due to the incorporation of some Trent technology in the high-pressure turbine blades and the intermediate compressor stator and rotor, and a new power turbine.

The Trent 60 DLE has the capability to drive its power turbine at either synchronous speed, so does not require a gear box to drive any generator. In the 50 Hz version its output is 51.5 MW with a gross heat rate of 8550 kJ/kWh. Exhaust flow rate is 151 kg/s at 444°C. The Trent 60 DLW with water injection has an output of 64 MW and a gross heat rate of 8755 kJ/kWh. Exhaust conditions are 171.2 kg/s at 408°C.

These large aero-derivatives have been mainly used for pipeline compressor drive and as utility peaking units. The advantage here is their rapid starting capability and relatively high efficiency. But in Indonesia there is the only combined cycle generating plant powered by aero-derivative gas turbines. The 66 MW combined cycle is located some 20 km north of Samurinda in East Kalimantan, and went into operation in 1999.

Until 1990 the combined cycle had been regarded largely as a superior base load system. But the greatest experience of combined cycle was in the Far East: Japan, Taiwan, Thailand and Indonesia. Meanwhile in the United States, the emphasis on gas turbine application had been industrial combined heat and power schemes. The Public Utilities Regulatory Powers Act (PURPA) of 1979 set up the conditions for industries to generate their own power and process heat, and sell surplus electricity to the public utility at a fair market price.

Deregulation of the British Electricity Supply Industry in 1990 came at a time when much more gas had become available to Europe from the North Sea and later from North Africa and Russia. Some combined cycles were built to replace old steam plants, but also for industrial combined heat and power schemes. Other countries followed Britain with deregulation of their electricity supply which opened the door to industrial combined heat and power.

The next big combined cycle development was the heat recovery boiler. As gas turbines have become bigger with, larger mass flow and higher exhaust temperatures, so the steam cycles have followed with higher pressures. For a traditional drum type evaporator that inevitably means thicker material for the drum and piping. In fact the drum boiler is at the limit of its performance and we need a boiler that can heat up quicker and bring the plant up to full power it has to be a once-through design.

Deregulation of electricity supply particularly in the United States has brought with it the rise of the Merchant Power plant. Nearly all the combined cycle power plants recently ordered are of this type, which have no long term power purchase agreement with a single customer, but trade their output on the spot market according to demand and their production costs and availability.

Therefore a Merchant plant must be able to start quickly and be available to generate at any time. So with even larger gas turbines coming into service the heat recovery boiler is a critical issue.

The origin of the Benson boiler goes back to the dawn of power engineering. A young German engineer emigrated to the United States before the First World War, and took the name Mark Benson. Having failed to get a US patent, he filed a second application in Germany in 1922 which was successful. Benson's idea was to heat water at a supercritical pressure to above the equivalent saturation temperature at which there is no discernable phase change. Opening a flow control valve would result in the discharge of dry steam at subcritical pressure. The idea led to a high pressure once-through boiler, the main feature of which is that there is no drum.

At this time Siemens, as a builder of generators and steam turbines, had ideas of adding boilers to their production and thereby be able to design and build the complete power station. In 1924 they acquired patent rights for the Benson boiler and built their first, with an output of 30 t/h, in 1926. In the event they only built three Benson boilers before they decided instead, in 1933, to license the design to boiler makers in Europe and North America. They continued however with research to support thc industry, and retained the name Benson which they used as a trademark.

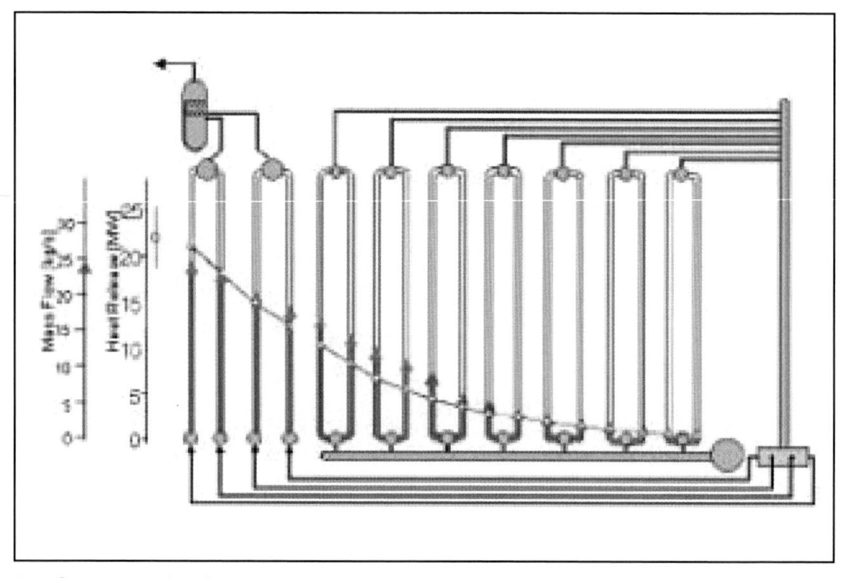

3.9 Schematic of Siemens' Benson boiler which is designed to fit in place of the high pressure evaporator superheater stages of a hotizontal drum-type heat recovery boiler. (Diagram courtesy of Siemens)

The Benson heat recovery boiler is a completely separate line of development which follows the application of flow characteristics of a natural circulation boiler to a once-through boiler with low mass flux in a horizontal gas path. This is the basic principle of the Cottam boiler as shown in Figure 3.1. There is no circulation as in a drum type boiler. The mass flow is upwards and a water/steam mixture flows from the upper headers to the vertical columns at the right and is sent on to the four bottom headers which supply the tube rows of the second stage which feed slightly superheated dry steam into the bottle, from where it continues to a conventional superheater.

Water chemistry is different from that for the drum-type heat recovery boilers. The deaerating condenser is followed by a two stage condensate extraction pump with a condensate polishing plant in side flow. The condensate polisher removes residual ammonia and oxygen which have escaped the condensate extraction system and the individual pressure flows are separately dosed with high ammonia and low oxygen in the drum stages and low ammonia and high oxygen (pH 8.5 and 50-150 ppb) in the Benson section.

Siemens are now using Benson boilers behind all their large

3.10 Cottam, UK: Siemens' 400 MW combined cycle which has the prototype Benson heat recovery boiler. In the background, EdF Energy's 2000 MW coal-fired power station, which entered service in 1969.

gas turbines. The first installation was in the United Kingdom at Cottam, near Nottingham in 1998. This was a single shaft system with the smaller SGT5-4000F rated 270 MW and a 130MW steam turbine. Now owned and operated by E.ON (UK) it logs about 150 starts per year.

A typical hot start after an overnight shut down starts by accelerating the gas turbine by driving the generator through the static frequency converter (SFC) only. Speed is then further increased with the GT start-up program by controlled fuel feed through the fuel control valve. From about 2000 rev/min the SFC is switched off and speed is increased to 3000 rev/min when the generator is synchronised and connected with the grid and the gas turbine is loaded to minimum output. Then it is ramped up to about 50% of full load according to the limits of the boiler.

The GT is held at this load, until the steam turbine is rolled off, accelerated and loaded to approximately 70% full load. When the bypass stations have closed the gas turbine is ramped up further until the required unit output is reached.

The gas turbine is not limited by the boiler. Since there is no high pressure drum the thermal inertia of the boiler is so much

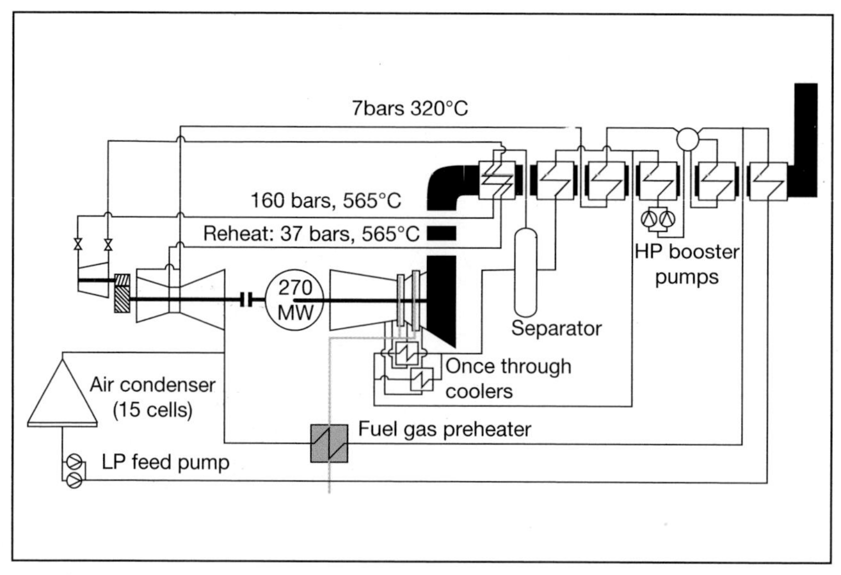

3.11 Schematic of two-pressure steam cycle designed by Alstom for their North American Merchant plants. In total 31 were constructed for GT24 single shaft blocks in the United States and Mexico.

lower and the gas turbine does not have to wait at part load for it to heat up. Gas turbine emissions are significantly reduced in the start-up phase because the NOx and CO emissions from a gas turbine are at their lowest above 50% of full load.

Cottam combined cycle went into operation in 1997 at about the same time as Alstom were repowering a 120 MW steam turbine at Badenwerke's Rheinhafen station in Karlsruhe with their prototype GT26. However the steam pressure at 156 bars was considered to be too high for a drum-type boiler, and instead a once-through boiler was used, which with no drum and therefore a smaller water volume, can heat up much quicker behind the gas turbine.

As a result of this, the company was approached by American National Power (ANP) to design a standard single-shaft power plant for the 60 Hz GT24, which could be a merchant plant in the deregulated American market.

With a merchant plant there is no long term sales contract to a specific customer: instead, power is bid on the spot market 24 hours ahead. Since this depends on there being high availability to exploit the best market opportunities, ANP wanted high part load efficiency and fast starting capability. The result was a 260 MW

3.12 Midlothian, TX: one of six 270 MW units of the GT24 combined cycle on site showing the 2-pressure boiler with once through HP stage and in the background the 15-cell air condenser.

single-shaft block with a 2-pressure boiler with a once-through high-pressure section at 160 bars, 565°C. In all, 31 examples have been built but as yet no equivalent version with GT26, which has been significantly upgraded since the design was first produced.

A feature of the GT24 combined cycles is the two-speed steam turbine. The full benefit of higher steam conditions can only be met with a more efficient steam turbine. Thirty years ago Stal Laval Turbin AB in Finspong, Sweden, who are now incorporated in Siemen's Industrial Turbomachinery Division, were a major supplier of steam turbines and epicyclic gear drives for the large merchant ships which were then being built. When that market collapsed, the company took stock of their steam turbine experience and applied it to designing a range of industrial steam turbines based on a series of standard modules.

This was the VAX range which was characterised by having a 9-stage HP turbine running at 8900 rev/min and connected through a gearbox to the LP turbine running at 3600 rev/min. In this arrangement, a combined IP and single-flow LP cylinder rotates at 3600 rev/min. At the exhaust end it is linked through the clutch to the generator, while at the reheat inlet end it is driven through the gearbox from the HP cylinder on the end of the line. The LP cylinder is therefore installed nearest to the generator and has a side-mounted condenser.

Among the first schemes in the United States to use this turbine was Richmond Cogeneration a 270 MW combined cycle in Virginia with two GT11N gas turbines and a 90 MW VAX turbine completed in 1991. Four years later Air Products installed a single-shaft combined cycle for a combined heat and power scheme in Orlando, Florida. This was a single-shaft block with GT11N and a 50 MW VAX steam turbine which went into operation in 1995.

The company was already looking at the combined cycle market for the VAX turbine and in June 1999 installed a 160 MW single-shaft combined cycle at Dighton, MA. This plant had the more powerful GT11N2 and a 50 MW steam turbine. Dighton, and the Orlando unit before it, can be considered to be the test plants preparing for the GT24 single-shaft units to follow.

The United States were the last to deregulate and since the previous utility structure was mainly state by state, that was how it was carried out. The New England States, Rhode Island, Connecticut, Massachusetts, New Hampshire, and Maine were among the first to deregulate and it is here that Alstom has dominated the market with their GT 24 combined cycles.

The first unit in service was not with ANP, but with Berkshire Power at Agawam, MA. This was the first of 21 blocks in the United States of which 11 are in New England and 10 in Texas. All had the same power train and heat recovery boiler. Agawam and two other sites have wet mechanical draught cooling systems: all the others have air condensers.

Another eight have been sold in Mexico, all at Monterrey, where CFE have a station with four blocks, and the Spanish utility Iberdrola has four in a neighbouring plant. Both plants have wet mechanical draught cooling schemes.

Air condensers have planning advantages; but disadvantages in performance because of the high auxiliary load when all fans are running. In planning, an air condenser avoids the issue of water abstraction for cooling which may rule out some sites for future developments.

Efficiency has a direct bearing on production cost. To achieve marginal gains from the steam cycle requires more than just higher starting steam conditions. For GT24 and the larger 50 Hz model GT26, the 32:1 compression ratio means that air bled from the compressor for turbine blade and vane cooling must itself first be cooled, and this heat is recovered for steam production. Once-through coolers connected between the final HP economiser outlet and the HP superheater output recover heat from this cooling air to boost the energy to the steam turbine.

Agawam when completed as the first of the GT24 combined cycles, had the highest steam conditions yet achieved in a commercially operating plant. The gas turbine exhaust flow into the boiler was 391 kg/s at 640°C which resulted in 160 bars, 565°C in the once-through HP stage, with the LP at 7 bars 320°C produced in a drum type evaporator.

By using the GT24, Berkshire Power can take advantage of the high efficiency which is characteristic of sequential combustion. Base load efficiency is 57.6% and at 50% of full load it is still 52%. An intake air cooler is powered by two gas-fired absorption chillers which supply chilled water to piping in front of the air intake filters. These come into operation as the ambient temperature reaches 15°C, to increase the gas turbine output during the peak summer months when the temperature is above 30°C.

The combination of the F-class steam cycle, in which steam conditions are similar to those of many sub-critical steam plants, together with the Benson boiler has in fifty years almost doubled the efficiency of power generation. Agawam could have had a marginally higher efficiency if it had direct cooling from a river. But to get to 60% efficiency is really an extension of the steam condition and the production of more powerful gas turbines with higher exhaust conditions.

After 1995, the F-class gas turbines following a series of upgrades by each manufacturer achieved combined cycle efficiencies as high as 58.5% it was natural to ask whether 60% efficiency would be possible. In Germany one F-class combined cycle has come very close at 59.7%. At Irsching about 100 km northeast of Munich unit 5 at E.ON's power station is an 860 MW combined cycle with two SGT5 4000 F gas turbines exhausting into Benson boilers supplying a 240 MW steam turbine. The unit

3.13 Irsching 5, Germany: probably the most efficient combined cycle with F-class gas turbines at 59.7%. Benson boilers for rapid starting with up to 250 starts/year. (Photo courtesy of Siemens)

was designed for load following with up to 250 starts a year.

The gas turbines have been progressively updated since their introduction in 1995 at 240 MW. Upgrades have taken the output from 258 MW in 2001 to the current output of 292 MW with an exhaust rating of 692 kg/s at 577 °C. Three factors contribute to the high efficiency: a high-powered gas turbine, Benson boilers, and direct cooling of the steam cycle from the River Danube.

For the immediate future a large number of combined cycles may be ordered in Europe to ensure sufficient capacity margin remains after LCPD is enforced in 2015. A big program is under way in the UK which expects to see all but one of its coal-fired plants shut down by 2020 and all but one of the currently operating nuclear plants by 2025.

The plants are being built on the sites of old power stations. Drakelow, Pembroke Willington and Carrington have all been demolished and the oil-fired plant on the Isle of Grain is among those to be shut down in 2015.

As the efficiency of the F-class combined cycles steadily increased towards 60%, the question was asked if efficiency could pass 60%. It also has a psychological significance because

3.14 Baglan Bay, UK: 480 MW combined cycle with prototype GE Frame 9H gas turbine. Following extensive testing, at GE factory and on site it went into commercial operation in September 2003. (Photo courtesy of General Electric)

it represents a doubling of thermal efficiency in little more than seventy years.

The launch of the last F-class gas turbines in the mid 1990s had hardly finished before the first of a new generation of gas turbines appeared. At the end of 1995 GE launched their largest gas turbine yet; at 330 MW the Frame 9H is designed for the 50 Hz market to power a single-shaft combined cycle with an efficiency of more than 60%.

GE has from the beginning designed gas turbines like the early aero engines with a ring of up to twelve combustion chambers arranged around the shaft. As firing temperatures increased so greater cooling had to be supplied to the combustion cans and transition ducts leading to the turbine stages. The solution was steam cooling of the transition ducts and of the first two of four turbine stages, on both stationary and rotating blades.

The challenge for GE therefore was to design not only steam cooled stationary components, but also rotating blades, and then integrating the cooling system with the steam cycle. From this it can be seen that Frame 9H was designed as the driver for a more powerful base load unit.

The chosen site for the prototype was in south Wales at Baglan Bay about 20 km east of Swansea where it arrived at the end of 2001 after an extensive programme of tests of components as well as the complete gas turbine. GE Were taking no chances after the problems they had with the prototype F-class machines. At Baglan Bay it was built into a 480 MW combined cycle and was fired for the first time in November 2002. A further ten months of testing the combined cycle followed before the station was officially opened by the Secretary of State for Wales in September 2003.

By the time Baglan Bay started commercial operation concerns about Global Warming had taken hold and across Europe energy policies had evolved to include a percentage of renewables in the future plant mix. If anything Governments saw renewables as more important than reducing the emissions and improving the efficiency of generating plant.

Siemens adopted a different design approach for their Model SGT5 8000H, announced in 2001, on the basis of what the future role of the combined cycle should be. They saw the possibility of the doubling of electricity demand by 2020 in some of their traditional markets in Asia, the Middle East and South America.

Also, with the large and growing commitment to renewable energy schemes which do not operate continuously the combined cycle should not primarily be a high efficiency base load plant but have great flexibility of operation so that it could back up other generating systems. It should therefore be designed for a large number of starts per year and always be used with a Benson boiler to ensure fast start-up.

Since 1980 the requirements for new power plants have been changing. Initially it was economic. The combined cycle was a more efficient plant which therefore offered lower fuel consumption, and lower emissions, and could be built quickly in stages. In the growing markets of south east Asia it was not uncommon to install the gas turbines first and run them for up to a year in simple cycle, while the heat recovery boilers and steam turbine were erected.

The introduction of the single shaft block in the 1990s was accompanied by the first concerns over global warming. The primary concern was not so much economics as emissions and now at the start of the 21st century the particular wish is for high efficiency and flexibility of operation.

With now so much renewable capacity in Europe and North America the need is for power plants that can start up and shut down quickly. These have been the guiding principles in Siemens' development of the SGT5 8000 H; an efficient gas turbine for combined cycle duty with a target efficiency of 60% and capable of putting 0.5 GW on line in half an hour.

The gas turbine is Siemens first design with a can-annular combustion system of which there are sixteen cans in the 50 Hz model here, and 12 in the smaller 60 Hz model SGT6 8000 H. With no water cooling to be introduced as the gas turbine warms up, the rapid starting time is retained. But there are other equally important issues concerning efficiency at part load operation.

In 2006 the gas turbine was tested at E.ON Kraftwerke's Irsching site in Bavaria, where the following two years of simple cycle test operation validated the design. SGT5 8000 H is rated at 375 MW which makes it the most powerful gas turbine currently available, and has an efficiency of 40%. The pressure ratio is 19.2 and the exhaust flow is 820 kg/s at 625°C.

Now it has been converted to a combined cycle with the addition of a Benson boiler and a 205 MW steam turbine. The net heat rate is quoted as 6000 kj/kWh giving an efficiency in excess of 60%. In fact it was announced in May 2011 that the 578 MW combined cycle on test over the previous 18 months had achieved efficiency of 60.75% which, for the time being, must be a world record for electricity generation. Furthermore, thanks to the Benson boiler 500 MW can come on line in 30 minutes after an overnight shut down.

Having now got their gas turbines running and marketable for combined cycle in the 50 Hz market, both companies are developing the 60 Hz models by scaling down from the existing designs. GE have already got their MS7001H in production and the first two are being installed on a site near Riverside, CA.

Siemens have their first six SGT6 8000 H under construction in Orlando, FL. The design is a scaled version of the 50 Hz machine with an initial rating of 274 MW at an efficiency of 40%. The main difference from the 50 Hz specification is the slightly higher

3.15 Irsching, Germany: showing Unit 4 a 540 MW single-shaft combined cycle with the prototype SGT5 8000H at 375 MW and efficiency of 60.75%. (Photo courtesy of Siemens)

pressure ration of 20:1. Exhaust output is 600 kg/s at 620°C.

In a single-shaft combined cycle block the rating is 410 MW with an efficiency of more than 60%. As with all European installations there will be a Benson heat recovery boiler. A warm start after an overnight shutdown is expected to take about 40 minutes. Clearly Siemens are promoting in both markets flexible operation at high efficiency.

First orders were announced in April 2010 by Florida Power and Light who are redeveloping their Cape Canaveral and Riviera Beach power plants each with three 410 MW single-shaft combined cycle blocks. Both stations were shut down in 2010 and demolished. It is planned to have the rebuilt plants completed and returned to service in 2013 and 2014 respectively.

Two other proposals for high efficiency combined cycles were announced in 2011 which are based on uprated F-class gas turbines, also for the 50 Hz market, and aiming for an efficiency of 61%. General Electric propose a scheme with their Frame 9FB, which is their most powerful air-cooled gas turbine, and which has been uprated from 284 MW to 338 MW with an efficiency of 40%.

This gas turbine was initially developed for the IGCC market,

3.16 Model showing arrangement of GE Flex-efficient 50 single-shaft combined cycle designed for flexible operation at 61% efficiency. Available in 2015. (Photo courtesy of General Electric)

but is now at the heart of a 510 MW combined cycle designated as model Flex Efficient 50, which is forecast to have an efficiency of more than 61%. The gas turbine's higher output follows the substitution of a new compressor and turbine stages.

The model in figure 3.16 above, shows the basic layout of the plant with a 182 MW steam turbine on the end of the line which appears to have a side-mounted condenser connected to either wet or hybrid mechanical draft cooling towers. First customer plants will be available in 2015.

Alstom have also made a major upgrade of their GT24 and GT26 models, rather than bring out new models for future combined cycle application. That may come later, but they also aim for 61% efficiency with the upgraded models. The GT24 output has been raised to 230 MW and the GT26 to 332 MW, again this is the result of a comprehensive reworking of the compressor and turbine stages of each machine. The compressor has 22 stages but a larger diameter to accommodate the higher mass flow. The number of variable inlet guide vanes is increased to four. The power turbine has added a fourth stage.

These changes have created gas turbines with higher part-load efficiency and the ability to park at 10% of output. The underlying application philosophy is to be able to ramp up quickly to respond to loss of renewable output. For instance, when photovoltaic output drops at sunset which, in northern Europe, is a peak time for much of the year, or if the wind drops or a wind farm is taken off line, or there is a sudden breakdown in a large thermal plant.

GT26 is to be offered in a 500 MW single-shaft combined cycle with an efficiency of 61%. For the 60 Hz market GT 24 is offered in a 2+2+1 arrangement giving a combined cycle of 700 MW at 61% efficiency.

Sequential combustion has always produced higher part-load efficiency and this has been enhanced in the new designs which can go down to a low parking load and then come up again in 30 minutes. But this is basically true of all the high efficiency combined cycles given the way they will be used in the future; high part load efficiency and rapid ramp up to meet sudden loss of capacity.

We may see these systems come into service sooner as the LCPD takes effect. RWE Npower in the UK will after 2015 have only one coal fired station at Aberthaw, near Cardiff, a 1500 MW plant with FGD and other environmental measures. All their other stations will be combined cycles. They plan a new plant at Willington which will be a 2000 MW combined cycle with four 500 MW blocks. Clearly they want a more efficient plant which can initially run base load and later go into the load following mode. This plant is planned for service in 2015.

It will surely not be unique. Combined cycles have taken a further step in development to give better than 60% efficiency and greater flexibility of operation. This will give lower emissions and greater security of supply in a system with a large renewable capacity and greater nuclear base load.

4 Combined heat and power

Deregulation of the electricity supply industries around the world after 1990 kick-started the combined heat and power business where none had existed before. Once it became possible for anybody to generate electricity it became possible for any industry to install a gas turbine and heat recovery boiler to supply their electricity and process steam independently of the public utility, and moreover get a fair market price for any electricity sold which was surplus to requirements.

Typical industries for combined heat and power are pulp and paper, petrochemicals, oil-refineries, distilleries, food processing and the steel industry. Typical gas turbines in such applications are the large aero-derivatives and the small heavy frame units between 5 and 50 MW. The largest gas turbines running at synchronous speeds are used in specific jobs, for example as a single power plant supplying a large industrial estate with power and process steam, or in desert areas of the Middle East to distill sea water for the public water supply.

As a generator for combined heat and power the gas turbine has several advantages. The gas turbine exhaust volume and temperature determines how much steam or hot water can be supplied to process, and can be boosted by supplementary firing if necessary. But however much process steam is supplied from the heat recovery boiler, then the electrical output of the gas turbine is effectively constant.

Today some industries can produce some or all of their own fuel which further encourages them to invest in CHP. For example, black liquor from pulp mills, hydrogen from some chemical processes, and blast furnace gas from the steel industry. Gas turbines have

4.1 Askar, Bahrain: one of two 420 MW combined cycles at Aluminium Bahrain built in 1993 to support growth of production up to the present time. Output is currently almost 900,000 t/year.

been modified to burn hydrogen and blast furnace gas, and particularly the latter has been an important factor in the choice of combined heat and power schemes for the steel industry.

It was around the time of the first oil crisis in 1973 that the first gas turbines capable of running at synchronous speed appeared as designs to be used for electric power generation. Natural gas had by then been discovered in the North Sea but it had first to replace manufactured coal gas for cooking and space heating in the domestic market. In fact there was a shortage of gas for power generation across much of Europe until about 1989-90.

Before then it seemed that the very countries which had pushed up the price of oil and gas were now telling us how to use it efficiently. The states along the gulf coast of the Arabian Peninsula have developed through their oil and gas resources and in the last thirty years the population has grown rapidly with the influx of Asian workers and Western and Japanese businessmen to build and operate the new infrastructure. Kuwait, Bahrain, Qatar, the United Arab Emirates, Oman and Saudi Arabia all rely on sea-water distillation for their public water supply.

But there are industries such as aluminium production which

4.2 Dubai Aluminium: the original power plant which still supplies water to the smaller emirates, bottled water to restaurants and demineralised make up water for Dubai Electric's neighbouring Jebel Ali station.

use only electricity for the process and in the Middle East not only sell surplus power to the public utility but generate some of it in combined heat and power plants. The first aluminium refinery in the Middle East was in Bahrain, which started production in 1971. The power supply was by simple cycle gas turbines which were added up to a total of 24 as production increased. The last five of these gas turbines were converted into a combined cycle in 1988.

The following year ABB supplied six GT13D for a further extension and the decision was taken to replace the nineteen simple cycle units by converting the new gas turbines into two combined cycle blocks by adding six dual-pressure heat recovery boilers and two 120 MW steam turbines. The combined cycles in 3+3+1 arrangement went into full commercial operation in October 1993 whereupon the company negotiated a power sales agreement with the Department of Electricity to use the old gas turbines to supply up to 1270 GWh per year to the public power system. Today aluminium production capacity is about 870,000 t/year.

In 1977 the Ruler of Dubai decided to follow Bahrain and build an aluminium refinery together with a gas turbine power station to supply the electricity required. The plant at Jebel Ali went into operation in 1979 with initially three pot lines which have since been increased to six. Dubal's original power plant had twelve gas turbines, of which eight were GE Frame 5's and four Frame 9, each exhausting into a heat recovery boiler supplying steam to a multi-stage flash distillation unit.

The power plant no longer supplies water to Dubai City, but they do supply bottled water to restaurants and a regular procession of tanker trucks takes water to the smaller Emirates further north. Dubal produce high quality distilled water for boiler make-up and various industries in the Emirates. Public water supply for the city now comes from Dubai Electric's Jebel Ali power plants, which take boiler make-up water from Dubal's distilled water output.

Dubal now supply their smelter with combined cycles. When in 1996 they decided to install a new pot line, they bought two PG9171E gas turbines from GE Europe. Then in 1997 they decided to convert these to combined cycle and add two more gas turbines of the same type for a second combined cycle block to create a total capacity increase of 450 MW under site conditions and allow them to stand down some of the older gas turbines to reserve. This plant is known as Condor. Then in 2002 ABB received an order for the refurbishment of the first three pot lines and addition of more reducing cells to increase production capacity to 710,000 t/year.

To support the increased production the Kestrel power plant is a 330 MW combined cycle, similar to, and built alongside the Condor plant with two GE Frame 9E and a steam turbine which had to be built to be available at the end of the refinery extension, in the summer of 2004. Production in 2010 was 1,043,104 t/ of aluminium products.

The Jebel Ali power plants were built as combined cycles in a configuration that guarantees that demand for water is virtually independent of power demand. Similar plants have been built at Al Taweelah, Abu Dhabi, and on a new site at Al Shuweihat where two 1500 MW combined cycles with water production of 380,000 m^3/day came into operation in 2004 and 2011 respectively.

These dual purpose plants are designed so that water supply is independent of electricity supply. Three 280 MW gas turbines with fired heat recovery boilers supply steam to two turbines each supplying two multi-stage flash distillation units. The fired boilers each have a forced draught fan which allows them to operate as a fully fired boiler if the associated gas turbine goes down for any reason. Similarly, if the steam turbine is down the outputs of the boilers can be throttled down, bypassing the steam turbine, and connecting to the headers feeding the distillation units.

Operating in this way, water supply can be guaranteed all year round and therefore it is electricity demand which is the determining factor as to when the gas turbines will run. In the Emirates there is a low demand for electricity in the winter months and as soon as the hot humid weather creeps up the Gulf coast towards Kuwait, the air conditioning load builds up and electricity demand is almost tripled.

Since these plants were built a remarkable change has come over the landscape. Modern sewage works have been built to keep pace with water demand and the clean water effluent is used for irrigation. As a result, there is in Dubai a championship golf course, one of three in the Emirate, and extensive tree planting conceals the smelter and the Jebel Ali industrial zone from the main Abu Dhabi road.

This is a rather extreme consequence of a combined heat and power application. Elsewhere, supplementary firing can even out differences in process steam and electricity demand. All along the Gulf Coast the demand for water is met through combined heat and power schemes from the sea. Growth in public demand for water is what determines when and where a new power station will be built.

One other water and power CHP scheme in the region is at the Muscat Naval Base in Oman, which occupies a strategically important position at the entrance to the Gulf guarding one of the major oil and gas export routes from the Middle East. In 1983 the Government of Oman decided to install a combined heat and power scheme at the Base to provide its power and fresh water needs.

As designed water production has priority over power. The plant consists of two GT35 gas turbines from ASEA Stal (now Siemens Industrial Power Division) with single pressure heat recovery boilers together with an auxiliary boiler to guarantee steam supply to the distillation units when either gas turbine is down for maintenance. The heat recovery boilers have supplementary

4.3 Abu Dhabi, Al Taweelah: this desalination plant is one of several combined cycles in the UAE which supply steam to multi-stage flash distillers to create the public water supply. (Photo courtesy of Siemens)

firing with natural gas to ensure stability of water supply as the gas turbine outputs vary with temperature.

Several of these Middle Eastern CHP schemes were completed before 1990 when in the rest of the world there was still widespread resistance among electric utilities to combined heat and power. To bleed off steam to supply say a paper mill meant that there was less steam available to generate electricity and the price that the utility could get for the steam was less than it could get for the electricity.

Yet by 1990 there had been installed a large number of small combined heat and power schemes for specific industries in which the efficiency expressed in terms of useful energy out against fuel energy in was calculated at around 80 percent.

Most of these schemes were in the United States where in 1979, the Public Utilities Regulatory Powers Act (PURPA) was introduced by the Carter administration. Under this measure anybody could build a combined heat and power plant and would be guaranteed the market price for their electricity sales back to the grid. To qualify a PURPA plant had to have a dedicated steam host, in a large industry such as a paper mill, or a chemical plant.

4.4 Midland, MI, USA: uncompleted nuclear plant at right provided the steam turbine for the 1560 MW combined cycle sending 614 t/h of process steam to Dow Chemical. (Photo courtesy of Midland Cogeneration)

The fact that the United States was a country of over 200 million people with a largely uniform life style, meant that many industries were on a much larger scale with production plants in different regions, each serving a market bigger than that of many other countries in total for the same product. Thus if one factory in, say Texas, installed a combined heat and power scheme an enlightened management could install similar schemes in their other factories.

The larger American schemes would be combined cycle plants with one or more of the GE Frame 7EA or Westinghouse 501D or ABB GT11N, with most of the large installations in California, Texas, and the northeastern states. A number of schemes used large aero derivatives such as the GE LM2500 rated at 25 MW and the 33 MW LM 3000 which was later replaced by the 42 MW LM 6000. Many of the smaller schemes were simply a gas turbine and heat recovery boiler supplying steam to an existing turbine.

The largest American scheme at Midland, MI, began as a nuclear CHP scheme to serve the Dow Chemical works. When the power plant was cancelled in 1984 it was about 85% complete. It was later decided to modify one of the steam turbines to accept the output from the heat recovery boilers of twelve ABB Type

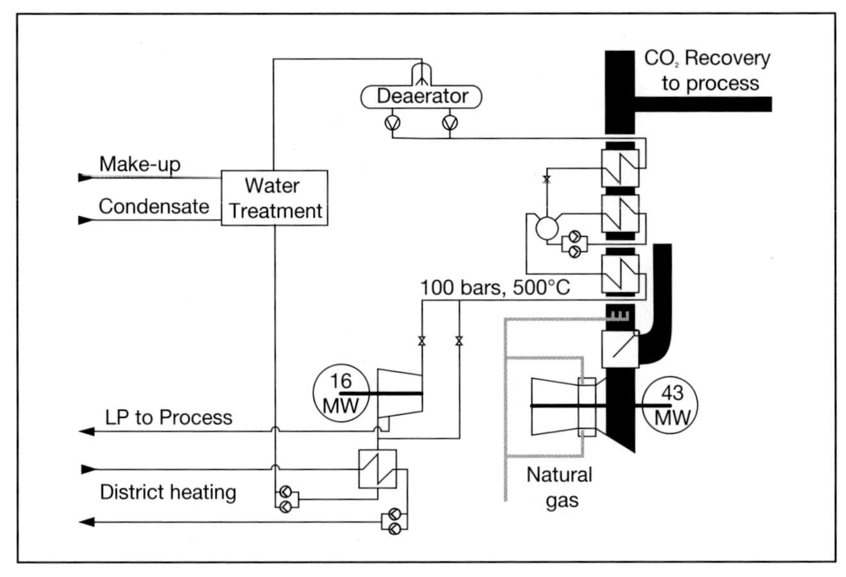

4.5 Aalost, Belgium: Amylum NV built this combined heat and power scheme 1996. Carbon dioxide recovered from the stack is used to make fizzy drinks and a bleed from the steam turbine supplies the town's district heating system.

GT11N gas turbines. The converted plant went into service in 1991 with outputs of 1560 MW and 614 t/h of process steam to Dow Chemical. In 2009 Midland Cogeneration Venture was taken over by EQT Investments with 70% and Fortistar with 30%.

The opportunities for combined heat and power over much of the world did not appear until after deregulation of electricity supply, starting in the United Kingdom in 1990, which effectively separated generation from transmission and distribution and allowed independent power producers into the market. This was only possible because the enabling legislation also defined the terms of power sales contracts.

The new generating companies quickly established project engineering divisions to develop CHP schemes. A typical scheme could be on a plot of land outside, say, a paper mill. The mill had old boiler plant which it wanted to replace and was intending to install a new paper machine. The generating company would design a power plant which would export steam over the fence to the paper mill, but feed the power into its network which they would then sell back to the industry.

Several such schemes were built, some of them combined

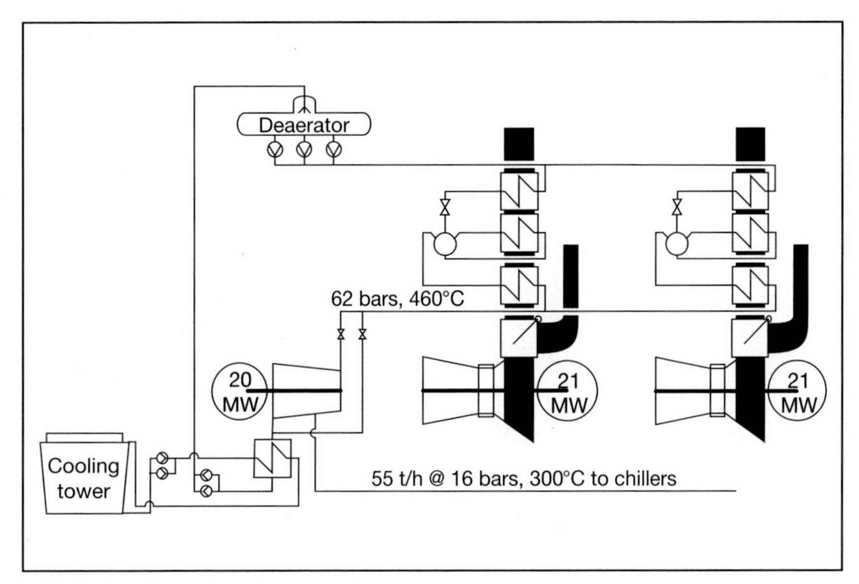

4.6 Bangkok International Airport, Thailand: this 60 MW combined heat and power scheme sends steam to absorption chillers for air conditioning in the terminal and the airline catering building.

cycles incorporating new or existing steam turbines. Buying electricity from the grid rather than direct from the power plant provides additional security, since power plant maintenance is covered by the power company who can continue to supply from another source. It depends on the importance of having continuous electricity supply to the process. Therefore the heat recovery boiler would have supplementary firing and be fitted with a forced-draught fan to cover steam supply as a fired boiler of the same output when turbomachinery is down.

Amylum NV manufacture starch and food additives and soft drinks at Aalst, about 60 km northeast of Brussels. In 1995 the company decided to replace old coal-fired boilers with a gas fired combined heat and power scheme. In addition to the company's power and process loads, a small district heating scheme in the town is also supplied from the plant. The scheme is based on a 42 MW aero-derivative gas turbine, LM6000, with a fired, single pressure heat recovery boiler feeding into the existing HP process line. A new 16 MW back-pressure turbine is fed from this line and supplies the district heating system. The power produced is sold to Electrabel and Amylum buy back what they need.

4.7 Bang Pakong, Thailand: Amata ECGO plant with two Siemens SGT 1000F gas turbines in a 180 MW combined cycle which supplies electricity and process steam to a developing industrial estate 70 km south of Bangkok.

Electrabel, and Tractebel Engineering are now a wholly owned subsidiaries of the French multinational Gulf Suez who have interests in combined heat and power schemes in Europe, Asia, and North America. Two of their largest schemes are in Thailand and Singapore. Glow Energy is the GDF Suez subsidiary in Thailand with two combined heat and power plants supplying industries in Mab Ta Phut, about 120 km south of Bangkok.

Deregulation in Thailand retained the Grid in public ownership so that the Electricity Generating Authority (EGAT) would eventually give up generation and buy all its power from independent producers. The Small Power Producers Act of 1992 defined such companies as operators of industrial CHP schemes who would sell a part of their output to EGAT under a long term contract.

The first small power producer to come into service was the Cogeneration Company of Thailand (CoCo) which at the time was owned by coal miner Ban Pu plc with 72% and a partnership of four Scandinavian utilities and IVO Holdings with the remaining 28%. The company was bought out by Tractebel in 2000.

The plant was designed as two 145 MW combined cycle blocks

4.8 SembCorp Cogen, Singapore: 635 MW combined cycle CHP scheme has dual pressure steam cycle for fast start up and supplementary firing to 780°C so that steam demand is independent of electricity demand.

each with three 35 MW, GE MS6001 B gas turbines and a 50 MW steam turbine. Each boiler at full load produces 70 t/h of steam at 52 bars 426°C some process steam is taken from the high-pressure output of the boilers, and a low pressure supply at 19 bars, 205°C is taken from the steam turbine. Total flow to process is 145 t/h, and condensate return is about 50%. The rest of the 19 bar steam is injected into the gas turbines for NOx reduction. 90 MW of the output of each block is sold to EGAT under a 21 year contract.

In Singapore, no combined heat and power was possible until, in January 1999, a contract for gas was signed with Indonesia for the West Natuna field offshore Sarawak. Ten years earlier a limited supply was negotiated with Malaysia which was burnt at the Senoko power station on the north shore of the island. The new 640 km-long pipeline was completed and brought into operation in January 2001.

The Singapore terminal is on Jurong Island which is the centre of the refining and petrochemical industries. A few schemes have been built with the F-class machines either serving large industries such as oil refineries or chemical plants, or else a cluster of smaller industries concentrated on an industrial estate.

Sembcorp Cogeneration is an example of a large combined heat and power scheme serving a whole industrial estate. The project is located at Sakra on a 2 hA site within a 3000 hA land reclamation which has created a centre for the petrochemical industry on the southwest corner of the island. SUT Sakra owns a large, above-ground pipe network which supplies process steam at three pressures, from its own boiler plant, along with high-grade industrial water, chilled water, sea water for cooling, and demineralised water to the petrochemical and chemical plants in the area. With the reorganization of electricity supply and the decision to import gas from Indonesia the CHP scheme was seen as a logical extension of their system.

In the deregulated Singapore market, power generators cannot engage in other utility services. Therefore SUT Sakra created SembCorp Cogeneration as a separate company to own and operate the power plant. The shareholders were SembCorp Utilities with 60%, Tractebel with 30%, and Economic Development Board Investments with 10%.

As a centrally controlled power plant, all of the electricity produced must be sold into the Singapore Power Pool so that it is dispatched on merit order of its electricity offer price and availability. The combined cycle therefore has been designed for high flexibility in operation so that its availability as an electricity generator is not compromised by the demands of process steam supply.

The plant has two GE Frame 9FA+e gas turbines and a 180 MW steam turbine from which the process steam supplies are taken. SUT Sakra's boiler plant remains on standby to cover maintenance of the gas turbines and steam turbine. With supplementary firing to 780°C, nearly 200°C above the natural exhaust temperature of the gas turbine, process steam demand can be followed independently of electricity demand.

The heat recovery boilers are a vertical, assisted-circulation design for higher reliability and availability for power generation and steam supply. Assisted circulation affords flexibility of operation with fast start-up and ease of repair in the event of leakage requiring the removal and replacement of individual boiler tubes. These boilers have a smaller footprint, which is important for a congested site such as on Jurong Island. Each boiler has its own deaerator and feed water tank mounted on the side of the frame.

Normally the gas turbines would power a tri-pressure steam cycle with reheat, but not here. The steam cycle is 2-pressure with reheat, which is best suited to the large process steam loads, but which also gives a boiler of lower thermal inertia which will be faster than a tri-pressure system to start up in electricity-only mode of operation.

In pure electricity mode, the plant has one of the highest pressures so far achieved in a drum type boiler at 145 bars, 550°C. Reheater outlet is at 41 bars, 518°C. With maximum process steam delivery, the high pressure output reduces to 104 bars, 566°C and the reheater to 24.6 bars, 566°C. Under this condition supplementary firing is at its maximum of 780°C.

The plant has been designed for ambient conditions of 27°C and 85% relative humidity, and a seawater temperature of 29°C. Under these conditions, and with maximum heat load, the electrical output is at 653 MW with an efficiency of over 60% when producing 550t/h of process steam. Steam outputs are carried on a pipe bridge which links with the SUT network at the plant boundary. There is a continuous supply of demineralised water coming back from SUT to top up the condensate return to the steam cycle.

The foregoing examples show that according to the energy outputs of the plant, it can be designed for priority as either an electricity producer or a steam producer. These are large schemes with F-class gas turbines. One very big scheme on Teesside in northeast England is one of the largest in the world with eight 133 MW gas turbines in two groups of four supplying a common steam range feeding two 400 MW steam turbines supplying process steam to the Wilton chemical works. The fuel supply was obtained by stopping flaring on a number of North Sea oil platforms and directing the gas into a collector main which fed into the British Gas system.

Teesside was shut down in April 2011 for economic reasons, leaving only the stand-by plant (an LM6000 aero-derivative with a heat recovery boiler) in operation. The present owners, Gulf Suez, have been considering rebuilding or selling it. But along the

4.9 Whitby, Canada: Atlantic Packaging board mill 60 km east of Toronto has the first Rolls-Royce Trent in a combined heat and power plant. Note once through heat recovery boiler. (Photo courtesy of Innovative Steam Technologies)

coast, near Redcar, E.ON are planning to build Thor Cogeneration a 1020 MW combined cycle which will also supply steam to the neighbouring North Tees Refinery.

A more important issue is the impending closure of several large coal- and oil-fired plants in 2015. The currently mothballed Teesside plant must be kept available as an electricity generator since on full load it produced 3% of the national electricity supply.

The large aero-derivative gas turbines, notably the Rolls-Royce RB211 and Trent, and the GE LM6000 have found a number of applications in combined heat and power schemes. The largest of these is the Trent, of which the first installation was in Canada at the Alantic Packaging board mill in Whitby, Ontario in 1998. A number of aero derivatives are installed in oil refineries, which are one of the important markets for combined heat and power. At least nine schemes were completed and put into operation between 1986 and 2005. In Europe the deregulated utilities set up design and construction teams to build some of these plants.

In Belgium, Electrabel supplied two refineries in Antwerp. First was at Esso's refinery site in the port area on the north side of the

4.10 Antwerp, Belgium: Fina Cogen is a 180 MW scheme with three LM6000 aero-derivatives. Plant was built by Electrobel to send steam over the fence to the refinery and electricity into the network to be sold back as needed.

Scheldt estuary. Separated oil fractions are then processed at the petrochemical plant on the same site. In 1990, the reorgnisation of the Belgian electricity supply system created opportunities for industrial CHP schemes either in full company ownership or in joint ventures with the electricity supply industry. Esso decided to expand their on site energy production with a combined heat and power scheme based on GE's Frame 6B gas turbine with a fired heat recovery boiler coupled into existing processes. The gas turbine would burn natural gas while supplementary firing could burn either natural gas or refinery waste gases. The plant was ordered in 1991 and went into full commercial operation in May 1993.

A much larger scheme is at the Fina Refinery where in 1997 they decided to replace an old boiler supplying a 60 MW steam turbine with a combined cycle. they entered a joint venture with Electrabel whereby the utility would build a 180 MW CHP scheme on land provided by Fina. Steam is supplied to the refinery and Electrabel take all of the power produced. The plant has three LM 6000PD aero-derivative gas turbines with supplementary fired boilers. Waste refinery gases are supplied to the duct burners and make-up water is supplied from the existing refinery plant. Provision has been made for future installation of a by-pass stack and fan on one unit in order to provide greater operational flexibility under varying conditions of steam and power demand.

In 1986 the Egyptian Petrochemical Company installed a gas turbine CHP system to serve petrochemical process plants on a new industrial estate at EI Ameriya, some 40 km west of Alexandria. This first plant comprised a 36 MW ABB Type 9 gas turbine, unfired heat recovery boiler and a 37 MW condensing turbine, with a back-up fired boiler, it was designed with future expansion in mind but process capacity did not start to increase significantly until 1998. Until then the power plant had operated in island mode with just an emergency connection to the public grid, but as chemical production increased the company became a net importer of electricity.

The plant extension was designed to increase steam supply and stop power imports. The more powerful 52 MW GT8 gas turbine was used with an unfired heat recovery boiler which was connected to the existing steam turbine and process headers to form a 105 MW combined cycle. A new deaerator was linked to the old unit and operates as one in a common feed water system.

The break up of the Soviet Union into its separate republics opened Azerbaijan and the Caspian oil port of Baku to western technology. In Soviet times Azerenerji was responsible for electricity supply in the whole of Azerbaijan. At the Baku refinery there was an old CHP scheme with back-pressure turbines dating from 1965.

The new scheme is not a repowering as much as a means of separating steam and electricity demands. With the old scheme, if electricity demand fell, the steam output rose whether or not the refinery wanted it. So the new plant had to be a more flexible system and was to be two 60 MW gas turbines with heat recovery boilers and supplementary firing.

ABB already had the 52 MW GT8C and at the end of 1997 the uprated model GT8C2 was introduced with an output of 57.3 MW and having improved compressor blade profiles scaled down from the GT24. The single-pressure heat recovery boilers have supplementary firing and can produce up to 200 t/h of steam at

16 bars, 285°C. A further degree of flexibility is offered with the fitting of a diverter valve and a bypass stack on one of the units.

In 1999 Azerenerji having got a new refinery CHP scheme under construction also decided to renew their generating plant and decided to put a 400 MW combined cycle at Severnaya, 50 km north of Baku. The plant is a single-shaft block of the 270 MW Mitsubishi MW701F, site rated 250 MW at 30°C, with a tri-pressure reheat steam cycle with a heat recovery boiler by Cockerill Mechanical Industries, of Belgium. The plant went into commercial operation in November 2002.

Back in Europe the Total Fina Elf refinery at Gonfreville, near Le Harvre, wanted to replace old boilers and COFIVA, a division of Electricité de France, suggested a combined heat and power scheme to supply the refinery with steam and power. The plant comprises two GE Frame 9E gas turbines and with supplementary firing in the heat recovery boilers. Steam production can be further boosted by using a forced draught fan in parallel with the gas turbine. If one gas turbine is shut down, then the fan operates alone with the fired boiler. The boilers are fired by the large volumes of waste gases produced in the processes. The power plant must guarantee steam demand to the refinery, depending on which there can be one or two gas turbines running with varying levels of supplementary firing. The plant was ordered in July 2001 and went into commercial operation in June 2004.

A similar sized scheme in Greece at the Thessalonica refinery has a single F-class gas turbine. Hellenic Petroleum has four refineries of which Thessalonica is the largest. Thessalonica Energy was created to build and operate the plant and is a partnership of Hellenic Petroleum, Tractebel, and AGEKEK, a Greek process engineering company. The plant was the first to be built after the deregulation of Greek electricity supply.

The gas turbine, a GE Frame 9FA+e exhausts into an unfired heat recovery boiler which supplies a 130 MW condensing steam turbine from which are extracted the three process pressures. Surplus power is sold to the Public Power Corporation.

Across the Atlantic Brazil is very much a hydro-powered country with two of the world's largest rivers, the Amazon and the Parana, flowing through it. These and their tributaries account for

4.11 Shanghai, China: Baoshan steelworks was the first to use the version of the GTN2 gas turbine modified to burn blast furnace gas in a combined cycle with a net output of 150 MW sent out.

more than 80% of the electricity produced in the country.

Industrial development and population growth have pushed up demand for electricity. However there is scope for combined heat and power, and Petrobras, the state owned oil company has installed 275 MW CHP schemes at its refineries in Salvador and Rio de Janeiro.

The first, known as Thermo Bahia, was installed at their Mataripe refinery near Salvador in January 2003. It was built by a partnership of Petrobras and ABB Energy Ventures. The gas turbine is a GT24 with a single pressure heat recovery boiler with heavy supplementary firing up to 794°C. At 25°C ambient electrical output is 240 MW with steam output to process at 387.4 t/h.

The second scheme, Thermo Rio, at the Reduc refinery in Rio was originally intended to be an identical plant to Thermo Bahia. However, Petrobras, this time in partnership with LG&E Energy of the United States and Sideco, the Italian steel maker decided on a 300 MW three shaft unit because Petrobras were also building on a neighbouring site two 300 MW combined cycle blocks each with two GT11N2 gas turbines. By changing the design to match the combined cycles, they had six identical gas turbines in the

4.12 Babrala, India: Tata Fertilizers run two Frame 5 with heat recovery boilers and a back-up fired boiler to provide their power and process steam in a stable environment independent of the grid.

same location with the advantage of commonality of spares. The only difference is the 80 MW back pressure turbine whereas the combined cycles have 100 MW condensing turbines.

The combined cycles were built in partnership with NRG Energy of the United States because of growing energy demand in south eastern Brazil which depended almost entirely on hydro power from Rio Parana and its tributaries. Output varied from year to year depending on rainfall, and was expected to do more in the future. There was therefore a case for efficient and fast starting capacity to complement the two base loaded nuclear plants at Angra dos Reis, approximately 170 km north of Sao Paulo.

GT11N2 has also been modified to burn blast furnace gas (BFG) with one installation in China and another in Japan. Baoshan Steelworks in Shanghai ordered their gas turbine in 1994 which was the first to be modified to burn blast furnace gas. The compressor uses the first fourteen stages of the GT24 compressor, including three variable inlet stages. The pressure ratio of 15.1 is the same as for the standard GT11N2 but the firing temperature is the same as for GT24 to raise the ISO base load rating from 109.5 to 115.4 MW.

The low calorific value of blast furnace gas, 2500 kJ/kg, compares with natural gas at 48,000kJ/kg, therefore the fuel mass flow is 140 kg/s compared with 8 kg/s for natural gas. The critical factor is mass flow through the compressor which has to be reduced with the inlet guide vanes to drop it from 375 kg/s in the standard gas turbine to 260 kg/s from the BFG fired unit. The combustor silo is also bigger and has a single diffusion burner with a 1.5 m diameter swirler modified from an existing BFD burner design.

The installation at Baoshan is a pure combined cycle on a single shaft. However China is a 50 Hz country and the gas turbine is a 60 Hz machine. The gas turbine is coupled to a Sulzer compressor which also runs at 3600 rev/min and raises the BFG pressure from 1.08 to 16 bars for the combustor. A reduction gearbox connects the compressor to the generator. Although the output of the BFG-fired unit is 144 MW the fuel gas compressor is rated at 51,5 MW so that Baoshan has a net output is 150 MW with 90 MW from the gas turbine and 60 MW from the steam turbine.

About nine months after the Baoshan combined cycle went into full commercial operation, at the end of 1998. Mizushima Steelworks in southern Japan ordered the same gas turbine for a combined heat and power plant which would be a combined cycle with steam take-off to power existing steam turbines. Mizushima is about 200 km southwest of Osaka on the 60 Hz network of Japan and it is therefore simpler since the gas turbine is effectively operating in simple cycle mode with a 2-pressure heat recovery boiler supplying process steam which among other things goes to steam turbines driving air compressors. There is no reduction gearbox to the generator which is in a unified power train with the gas turbine and fuel gas compressor.

Another country with a special requirement for combined heat and power is India. The grid system is inherently unstable because of the long distances between the power plants and the major load centres. Large variations of frequency occur from day to day and for an industrial plant with electrically driven pumps and compressors and processes controlled by timers, an unstable frequency can seriously affect production.

A number of industries decided that the only way to resolve this problem was to build their own power plant and run it in Island mode, independently from the grid. To be successful a plant had to be flexible in being able to operate power and steam supply separately.

Much of India's gas supply comes for the Mumbai High fields about 150 km offshore in the Indian Ocean. The fields were discovered in 1974 and production started in 1976. a pipeline was laid from the shore terminal north of Mumbai up to Delhi and terminates in the Tata Fertilizer plant at Babrala, 200 km further east. The company produces fertiliser from natural gas feedstock and has installed a 40 MW CHP plant with two MS5001 gas turbines, to supply power and process steam. The heat recovery boilers are single pressure with supplementary firing and are fitted with a second burner introduced below the LP evaporator in order to increase the steam produced at times of maximum demand.

Gas turbines each carry 10 MW of spinning reserve with 10 MW spinning reserve. Steam demand is followed on supplementary firing, and a separate fired boiler is a back-up. There are three process pressures, two of which are throttled down from the high pressure output of 112.5 bars, 515°C.

Some industrial power plants are set up to burn waste gases from the process. Quite often these gases are used for supplementary firing while the gas turbines themselves burn natural gas. Industries with byproduct gas available are oil refineries and large chemical plants. But in parts of Australia coal mines are an unusual source of gas turbine fuel.

About twenty years ago ABB supplied a 15 MW gas turbine to a coal mine in New South Wales. Methane extracted from the workings was compressed and supplied to the gas turbine which had a heat recovery boiler to provide hot water for pithead baths. Some similar schemes were installed in other countries where there are working coal mines with high methane content.

Queensland has extensive coal deposits and therefore depended on large coal-fired power plants for most of its electricity supply. But there are in the state some very deep deposits which it would be uneconomic to mine but which contain large quantities of methane. By recovering this gas and putting it into the natural gas pipeline system, gas turbine power plants and industrial schemes can burn this coal seam gas.

4.13 Mainz, Germany: Kraftwerk Mainz-Wiesbaden with backup plant at right, a fully-fired combined cycle dating from 1976, plants supply district heating and process steam to neighbouring chemical plant. (Photo courtesy of Siemens)

Bulmer Island combined heat and power scheme at the BP refinery near Brisbane was the first plant to take fuel from this new coal seam methane source. The power plant entered commercial operation in November 2004. In fact Australia is the only country in which the exploitation of coal-seam methane is a firm issue of energy policy.

There is one other application of combined heat and power which is widespread in northern and eastern Europe and Russia. District heating goes back to the start of the 20th century and in particular spread through the fuel importing countries with the rebuilding after the Second World War. But in the United States which pioneered it with steam systems, it has not developed much from modest beginnings. The original systems are still in use and are mainly in the downtown areas of northern cities serving a few large energy users for heating in winter and to drive absorption chillers or air conditioning in the summer.

District heating, as in many European cities, is a power plant supplying heat to a pressurized hot water mains system which distributes heat to public buildings and private dwellings around the city centre. The other is a similar but smaller system feeding a specific site such as a university campus or other large estate of public buildings, laboratories and workshops. Such schemes are seasonal in operation and are working in full combined mode for at most seven months of the year.

A large district heating scheme in Germany spanning the River Rhine is Mainz-Wiesbaden completed in 1999. The two towns face each other across the river, some 50 km north of Frankfurt. The plant is a 400 MW combined cycle with two SGT5 4000F gas turbines and a 120 MW, 3-cylinder steam turbine with bleeds to two district heating condensers. The output at 8 bars, leaves the plant at 130°C and returns at 70°C. Process steam is taken from the boiler IP at 15 bars. With no district heating load as in summer, the efficiency is 58%.

The first district heating schemes were based on back-pressure steam turbines. Some early schemes in North America and Europe used steam as the heating medium and were limited in the distance that they could transmit heat. Consequently district heating has never caught on in the United States but in Europe, particularly after the Second World War, district heating was extended with pressurised hot water as the medium, which was sent out at between 110 and 130°C and returning at between 50 and 70°C.

Many cities in northern Europe were held up as an example of what could be done with district heating, notably Stockholm, Västerås and Malmö, in Sweden and Helsinki in Finland, where coverage of the city centre and the surrounding housing areas was over 90%.

There are still a large number of back-pressure turbines in district heating schemes in eastern Europe and the CIS. Many of these are now being, or have been replaced by combined cycle plants with condensing/extraction turbines. In Europe, and particularly in Scandinavia and eastern Europe, district heating is widespread with large areas of domestic heating. These are all systems using presssurized hot water running through heat exchangers in the customer's premises. Typical temperature out from the power station would be 120°C with return at 60°C.

The reunification of Germany brought renewed combined heat and power to the Eastern part, specifically in the renewal of district heating plants in some of the major cities. These were coal fired plants with back-pressure turbines, some of them predating the Second World War. Dresden was the first city to adopt a combined cycle with two of Siemen's 65 MW SGT1000F gas turbines and a single condensing/extraction steam turbine. This was followed by new district heating schemes for Chemnitz, Leipzig and Halle.

The SGT1000F, at its current rating of 65 MW is the smallest of three scaled F-class gas turbine models introduced in 1995 and has been sold to a number of combined heat and power schemes around the world all in the form of a combined cycle with extraction from the steam turbine. The first commercial installation of these gas turbines was in a combined cycle for the Vuosaari district heating plant in the eastern suburbs of Helsinki.

The other small F-class machine is GE's Frame 6FA model derived from the 250 MW Frame 9FA. This, too has found application in large industrial combined heat and power schemes and small combined cycle plants. Both machines incorporate the technology of the big F-class machines particularly in the design of turbine and compressor blades and improved turbine cooling. Both have cold-end drive to the generator, and the dry, low emissions combustion system of the equivalent large gas turbines.

Because both turbines are scaled down from larger 3000 rev/min designs, under the rules of similarity they cannot run at a synchronous speed and must therefore drive the generator through a gearbox. Both gas turbines can therefore be applied to schemes in the 50 and 60 Hz markets. The rated speed for SGT1000F is 5400 rev/min, and for the GE Frame 6FA it is 5230 rev/min.

Several of the larger gas turbines of the E-class have also found niche markets in district heating and industrial CHP schemes. The main contenders here, for the 50 Hz market are the Alstom GT13E2, the GE Frame 9E, and Siemens SGT5 2000E all of which run at the 3000 rev/min synchronous speed.

The advantage of a combined cycle for district heating is that it can function as a high efficiency generating plant in the summer. It will have a maintenance outage of about three weeks before the start of the next heating season.

A number of large combined cycles have been installed for district heating Korea Electric Power Corporation, in which an SSS clutch has been installed between the intermediate and

4.14 St. Petersburg, Russia: the north St Petersburg district heating plant with two Siemens SGT5 2000E gas turbines which was the first of four 460 MW combined cycle blocks to support new housing developments.

low pressure cylinders. During the heating season the clutch is open and the low pressure cylinder is at standstill. The turbine is effectively operating as a back-pressure turbine with the high- and low-pressure cylinders. Then in the summer time the clutch is closed, the district heating condensers are shut off and the unit operates as a combined cycle with an efficiency of about 56%.

But there is also the future prospect of nuclear district heating. Forty years ago it was being considered in Sweden for Stockholm and Malmo and three other towns in the south. A period of anti-nuclear government prevented the plants being built and closed the Barsebac reactors which on a clear day were visible from Copenhagen, the capital of a staunchly anti-nuclear country.

However there was a way around this. Sweden, at that time, had an electricity supply system growing on nuclear supported by hydro. The only fossil-fired generation was in the district heating plants and industrial boilers. So why not reduce fuel consumption by using heat pumps to supplement district heating? These would of course provide a large electrical load for the nuclear power plants 24 hours a day.

Combined heat and power is truly a much greener energy

system which was largely neglected before deregulation of electricity supply at the end of the last century. Efficiency of such schemes, depending on the size of the heat load can range from the low sixties to nearly 90%.

Gas turbines account for the majority of industrial schemes built in the last thirty years and combined cycles are replacing back-pressure turbines in the old coal- and oil-fired district heating stations of the past.

Combined heat and power has expanded from modest beginnings following the deregulation of electricity supply around the world and most recent installations have used gas turbines as the prime mover. But the wider application of combined heat and power in all its forms can only benefit us in the long term with less air pollution and greater security of energy supply.

5 What future for coal?

Of all the known fuels, coal has occupied a special place in the countries that produce it. It made possible the industrial revolution as a fuel to produce steam to drive pumps, trains, ships, power plants and for domestic and industrial boilers. It is the most abundant of the fossil fuels, but the most difficult to extract, handle and store. It has killed many of the people who mined it and destroyed the health of most of the rest, and burning it has brought smog to the cities, and acid rain to lakes and rivers. In no way can it be described as a green energy source.

In the last fifty years coal has lost most of its traditional markets, leaving only power generation and parts of the steel and chemical industries as its principal uses. Now its days are numbered in the power generating market because all the effort to clean up its emissions, starting with flue gas desulphurization, has reduced the performance of the power station and increased the cost of construction and therefore of the electricity produced.

Of course where natural gas has become the substitute fuel, as in domestic and commercial space heating, the emissions are lower and contain a lot of water as a combustion product of methane. In the case of railway electrification much of the electricity is generated in coal-fired power stations.

After 1973 a serious effort was made to develop gas turbine applications with coal and with cleaner combustion. What resulted were practical solutions which have occupied niche markets and which have been overtaken by the decline of the European coal industries and the growth of natural gas in the energy markets around the world. Most of these projects have been technology demonstrations funded by national governments and the European

5.1 Schwarze Pumpe, Germany: this 1600 MW supercritical steam plant burns locally mined lignite, sends steam to a briquette factory, and in winter supplies district heating. At maximun heat load, efficiency is 55%.

Union and proof of technology has been more important than absolute performance.

Parallel with these developments materials technology advanced to make supercritical steam cycles economic and larger coal-fired plants were built with turbines of around 800 MW with steam conditions of say 260 bars, 560°C with reheat at 53 bars, 560°C. While a number of steam plants have been built with supercritical steam cycles which give efficiencies up to 45%. The vast majority of steam plants in the world run on a subcritical steam cycle at typically 150 bars, 540°C and reheat at 27 bars, 540°C, at an efficiency of at best 36%.

Drax, a 4000 MW coal-fired power station in South Yorkshire, is larger than the majority of coal fired stations but operates on a subcritical steam cycle at 166 bars 586°C, However it conforms to the LCPD condition having fitted flue gas desulphurization to all six boilers and also boosted over fire air technology to cut NOx emissions. Surprisingly these systems are absent from a lot of coal-fired plants in the United States of similar age.

At Drax a programme of turbine upgrades has raised the efficiency and a direct firing biomass system enables up to 10%

5.2 Drax, UK: this 4000 MW coal-fired power station was the last to be built in the UK and will probably be the last in operation. Meets LCPD condition and burns 10% biomass. (Photo courtesy of Drax Power Ltd)

to be fed into each boiler, bypassing the coal mills and burners. This is part of an ongoing programme to reduce greenhouse gas emissions. The biomass is a locally grown coppice willow supplied by Renewable Energy Growers, a company representing a group of farmers in Yorkshire and Lincolnshire who have planted some 1100 hA of willow and harvest about 300 hA per year. They supply biomass to Drax Power, and also to EdF Energy at Cottam, and a 30 MW wood-fired power station at Wilton on Teesside By 2025 Drax will have been operating for 40 years and now supplies 7% of the total British electricity supply.

If we are to continue using coal to generate electricity something has to be done about its emissions, including its ash. Concern for global warming has brought this about and the Green Movement is highly critical of any plans for new coal-fired power stations. It is conditional that any new plant must have carbon capture and storage which hasn't been economically proven on the scale required.

This will inevitably push up the price of coal firing and of the products that use the carbon dioxide that is recovered. So the price of energy in future will include the price of the power plant and its fuel and the disposal of its gaseous effluent. That if nothing else requires the construction of a pipeline to carry the gas to the user.

Coal and oil do not occur in the same place therefore if the powerplant sends carbon dioxide to a distant oil field who will pay for the pipeline, but enhanced oil recovery is getting more oil out of the well under the boosted pressure. There may be industries nearer to the powerplant which will have a use for carbon dioxide as a chemical reagent to make fertilizer or for the manufacture of effervescent drinks.

Such carbon capture systems as have been tested on existing plants have been seen to work but on a small scale, typically the flue gas equivalent to 10 MW of output of an 800 MW steam turbine. But it is not a new technology: carbon capture is a long-established chemical process and is often used in the oil and gas industry. Say a gas field has a percentage of carbon dioxide in it. This has to be removed from the gas to bring it up to the required standard for public supply.

But carbon capture is not just for coal firing. Combined cycles are targeted but they can already be seen to be a path to cleaner combustion. Furthermore if industrial CHP schemes are required to fit carbon capture there may be a use for the recovered gas in an industrial process.

Two processes have been developed offering cleaner coal combustion. The first was the coal-fired gas turbine developed in Sweden. The gas turbine does not have the regular combustor sytem but uses the hot exhaust gas from a pressurized-fluidised bed boiler. They produce no sulphur or nitrogen oxides under the conditions of combustion, and the Swedish plant has been used to test a carbon capture system with favourable results.

The other system is the Integrated Gasifier Combined Cycle (IGCC) which has a gasifier producing gas for a combined cycle which can supply compressed air to the gasifier process and take steam generated in the synthetic gas cooler.

Fluidised bed combustion was developed in the years after the Second World War and is inherently clean, since any sulphur in the coal can be chemically removed as it burns. But in the early days electric utilities were sceptical and wanted nothing to do with it.

The five plants are the most successful attempt yet to use coal as a fuel for gas turbines, and all are still in operation. It all started in 1974 with the publication by the then manager of ABB Stal's London office, Henrik Harboe, of a paper entitled *The Importance of Coal*. In it he proposed the use of a pressurized fluidized bed combustor in place of the seven can-annular combustors of his company's GT35 gas turbine.

The PFBC gas turbine, as it became known, was developed over a 20 year period with funding from IEA and the UK, German and US Governments. Development started in the UK and moved to Sweden in 1980. These plants have all been built as combined cycles and the steam cycle is shared between the fluidized bed and the gas turbine heat recovery boiler.

The first commercial application of a PFBC coal-fired gas turbine was to the Vartan district heating plant in Stockholm, and among the four other units that have been built are included a new district heating station at Cottbus, Germany, burning locally mined lignite, and a 390 MW combined cycle station on the Japanese island of Kyushu burning Australian bituminous coal. Two other plants at Escatron, Spain, and at Wakamatsu, Japan, are single GT35P which repowered old steam sets.

American Electric Power initially supported PFBC gas turbine and hosted one of the first three projects of the coal-fired gas turbine at their Tidd power station on the Ohio River west of Cincinatti. It was very much a research tool backed by DOE funding of $67 million, to study the clean combustion of various American bituminous coals with high sulphur and ash content such as the notorious Illinois No.5. The power station was 42 years old when the PFBC unit was commissioned, and over the next four years the properties of the system were thoroughly investigated and various filter systems were tested. It was closed in 1995.

No other plants have been built although interest is growing in the pressurized fluidized bed combustor which gives almost complete combustion, with no sulphur oxides emitted, and very low NOx because of the low firing temperature.

In a fluidised bed inert material, such as sand is kept in circulation by a current of air rising through it, As coal is injected into the bed it circulates with the sand as it burns. By lacing the bed with limestone any sulphur in the coal reacts with it to form the

5.3 Finspong, Sweden: the 75 MW GT140 PFBC gas turbine on test before it was shipped to Japan for Kyushu Electric's Karita project. It went into operation in 2001. (Photo courtesy of Siemens)

solid calcium sulphate which is removed with the ash.

The hot air rising off the bed contains dust which is collected in cyclones and returned to the bottom of the bed. It is a clean coal combustion system which produces an inert ash which can be used as an aggregate by the construction industry. Also the coal, although it has to be ground down to be injected into the bed it is not the fine powder produced by the pulverizing mills of a conventional power station boiler.

The basis of the coal-fired gas turbine was the GT35 gas turbine which had started life in the 1950's as a development program for an aero engine for the Swedish Air Force. The project was cancelled when the Air Force opted for the Rolls-Royce Avon but the developers, Stal Laval Turbin AB (now incorporated in Siemens Industrial Power Division) decided to adapt the engine for industrial application. The GT35 is still in production as the Siemens SGT500 and many units have been sold around the world for power plants, offshore platforms, district heating and industrial combined heat and power schemes.

The GT35P has little in common with the standard gas turbine. In place of the combustion chamber there is a fast acting shut-off

5.4 Karita Japan: panoramic view of the Karita power station with the PFBC unit at back, right which has been operating eleven years, and burning Australian bitumenous coal. (Photo courtesy of Kyushu Electric)

valve mounted on top of the gas turbine casing and leading into a coaxial pressure duct connecting to the fluidised bed combustor.

The gas turbine compressor is divided into high and low-pressure sections which are separated and mounted at opposite ends of the machine. The high pressure shaft is connected to the generator and drives it through a gearbox at 1500 rev/min. The low pressure section is a free turbine stage on a separate shaft driving the low pressure compressor. There is an intercooler between the compressor sections which is included to ensure that the temperature entering the PFBC is no more than 300°C.

The PFBC is a large cylindrical pressure vessel which is held at 12.3 bars together with the two cyclones which recycle dust coming off the top of the bed to the bottom. The type of coal being burnt defines the need for limestone sorbent for sulphur entrapment. Depending on the type of coal and its sulphur content the sorbent can be fed in dry or as a paste. At the bottom of the pressure vessel is a lock hopper which allows ash to be removed to hold the top of the bed at a constant level.

All of the PFBC plants have been designed as combined cycles, mostly by repowering old steam turbines. Of the three new plants, two supply district heating in Stockholm and Cottbus, Germany, and Karita is a 390 MW combined cycle generating plant owned and operated by Kyushu Electric in southern Japan.

The first PFBC gas turbines in service were at an extension of the Vartan district heating station in Stockholm. When Stockholm's Energi ordered the plant in the spring of 1987 it was at the culmination of 4000 hours of testing of the Process Test Facility in Malmö, a heavily instrumented, pressurized fluidized bed which was installed to determine the operating characteristics and do combustion tests on a number of coals, but mostly with Polish coals alone, and also when mixed with biomass fuels such as wood chips and palm nut shells. The fuel at Vartan is Polish coal.

The third GT35P was at Escatron, Spain, owned by the utility ENDESA the plant came into operation in 1991, a combined cycle of 63 MW of which the gas turbine contributes 17 MW the plant burns locally mined coal with 5% sulphur and 27% ash.

All the PFBC plants have been supplied as combined cycles, but only three as completely new power plants. Of these, Kyushu Electric's Karita power plant on the Japanese island of Kyushu is the largest and is interesting for the fact that a detailed economic evaluation at the time showed that the PFBC plant would have lower operating costs burning Australian coal than an equivalent sized combined cycle burning gasified LNG also imported from Australia.

Karita is the only example of the larger GT140P gas turbine, which is rated at 75 MW and integrated with the steam cycle of a 290 MW supercritical turbogenerator, for a total plant capacity of 365 MW. The GT140P was derived from the GT200, which was a large aero-derivative engine jointly developed in the 1970s by Pratt & Whitney and ABB. Only one was ever built and was installed as a simple-cycle peaking unit at Linköping for Vattenfall.

The pressurised fluidised bed in Karita is held at 16 bars. It is the only heat source and is not only the combustor for the gas turbine but also contains a once through heat exchanger which is the evaporator and superheater of the 241 bar steam supply to the main turbine. The plant went into commercial operation in 2001.

The GT 140 P followed the same general layout as the smaller GT35P. The 8-stage low-pressure, and 12-stage high pressure compressor sections were derived from the GT200 compressor, but the power turbine stages were scaled up from GT35. The operating temperatures are the same for both engines. The intercooler and the gas turbine heat recovery boiler are the feedwater heater and economiser for the main steam cycle which has the evaporator and superheater in the fluidised bed.

Of the bitumenous coals burnt at Karita, the average heating value is 26 MJ/kg with less than 1% sulphur, up to 29% ash and a maximum 7% of water. The efficiency at Karita is 44% which is about the same as the best currently obtainable with a conventional coal-fired steam plant on the same supercritical steam cycle, but of twice the size. complete with NOx catalyst, FGD system, and electrostatic precipitators. In effect this could be considered the definitive steam power plant for the 60 Hz networks.

The last PFBC station to be built was the new district heating plant for Stadtwerke Cottbus in eastern Germany. This has a single GT35P burning locally mined lignite. This brown coal has a heating value of 19 MJ/kg, contains less than 0.8% sulphur but 5.5% ash and 18% water.

The situation in 2011 is that the PFBC technology is looking for a market, at a time when interest in Europe and much of the rest of the world is still focused on gas-fired combined cycles.

PFBC is a proven clean coal technology which can handle a large range of coals from high grade bituminous to lignite with high ash and moisture content. However all the plants except Karita, Japan, have been subsidised by Governments, and most of them were built as demonstration units and were much smaller.

Karita is interesting because it is the only one of the plants which is of a comparable size to the alternative pulverized coal fired plants in the market place. Alone of the six, the efficiency is higher than that of the equivalent pulverized coal plant and at $1500/kW the cost is at the bottom end of the range currently predicted for pulverized coal-fired and IGCC plants. It is a low temperature system and so does not require a NOx catalyst, and the sorbent system replaces the flue gas desulfurization which in a coal-fired steam plant lowers the efficiency by putting a back-pressure on the boiler exhaust.

One other fluidised bed idea can be described as a modern

5.5 Map Ta Phut, Thailand: third power plant of Glow Energy Ltd is two PFBC boilers each with two GE Frame 6 gas turbines integrated into the feedwater heating system of a 170 MW steam turbine.

version of the fully-fired combined cycles built in the 1970s. This is in Thailand, 150 km south of Bangkok, at Map Ta Phut where it is one of three plants operated by Glow Energy Ltd, the Thai operation of Suez Tractabel, which supply power and process steam to industrial estates in the area, from three combined heat and power plants which send part of their electrical output on long term contracts to EGAT under the Small Power Producers Act.

The power plant consists of two blocks which are termed hybrid plants. Each consists of an atmospheric fluidized bed and two GE Frame 6B gas turbines with heat recovery boilers integrated into the steam cycle of a 170 MW steam turbine. It went into commercial operation in March 2000.

The Hybrid plant is a way of providing greater flexibility between electricity and process steam output. All process steam is generated in the fluidised bed and outputs to the two steam mains at 19 and 52 bars serving the Map Ta Phut estates. Glow Energy has two contracts with EGAT to supply 150 MW from each of the hybrid unitts: 30 MW from each gas turbine and 90 MW from the steam turbine. There is also a gas-fired combined cycle on the site which provides back-up power and steam to safeguard the EGAT

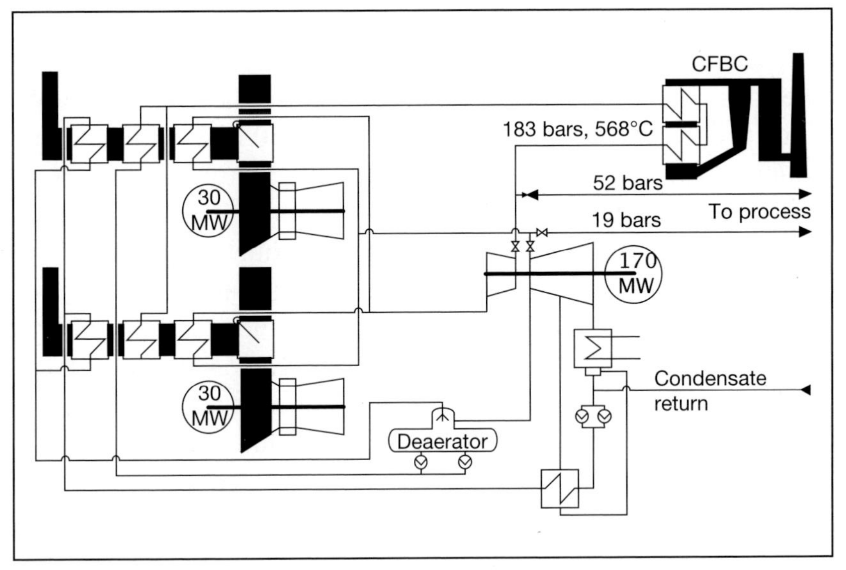

5.6 Map Ta Phut, Thailand: schematic of one of the combined cycle blocks at the Glow Energy plant 3. It is designed to safeguard EGAT's 150 MW power take off regardless of the demand for steam.

contracts during maintenance of the hybrid units.

But why a hybrid coal-and gas fired plant instead of another and larger gas-fired combined cycle? A combined cycle with heavy supplementary firing could probably have performed this duty, but under the Small Power Producer rules EGAT could only take up to 90 MW from each section of the plant and the remaining electricity demand has to be matched to the demand for process steam.

Compared with a conventional coal-fired plant the hybrid unit is inherently simpler. Without the gas turbines, the same capacity would be obtained with a 210 MW steam turbine with a reheater and various extractions to process, and with an array of maybe two LP and two HP feed heaters and a deaerator, all supplied from the energy of the coal. But there would not be the flexibility in balancing steam and power loads.

The hybrid system removes the reheater function from the fluidised bed boiler and puts it in the hottest region of the gas turbine heat recovery unit. All the energy of the coal is available to generate more steam, which is reheated by the gas turbine exhaust energy.

Thus the hybrid system is inherently flexible in that it can control the rate of combustion of coal to meet process steam demand while at the same time maintaining electricity supply to EGAT and the other customers. In fact the steam turbine set is rated 170 MW, which is enough to meet the EGAT load of 150 MW with both gas turbines down for maintenance.

The inert material of the fluidised bed is sand laced with finely ground limestone to absorb the sulphur in the coal. The combustion temperature is 800°C. The gas turbines however have steam injection for NOx control. Cyclones gather dust and unburnt carbon particles carried off in the flue gas stream, and returns them to the bottom of the bed. Any dust which is not returned through the cyclones is picked up in a chain of bag filters. from where it is collected, along with bottom ash from the bed and taken to a storage site from where it is sold as a construction aggregate.

The fluidised bed comprises once-through evaporator and superheater sections of the steam cycle which are at a much higher pressure than has so far been obtained in the largest combined cycles to date. Live steam output is 120.6 kg/s at 180 bars, 568°C with reheat at 530°C as determined by the exhaust conditions of the gas turbines.

As a combined cycle with no process steam load the output is 230 MW at an efficiency of 43%. With the full steam load of 100t/h to process the electrical output is 209 MW and the net cogeneration efficiency is 53%. Of the total energy output 55% can be attributed to coal and 45% to gas.

The Hybrid power plant is unique. It has been in operation since 2000 and has met and exceeded all performance guarantees. With the particular arrangements for power sales from independent cogeneration operators in Thailand it is particularly valuable in being able to meet steam demand without compromising its sales contracts to EGAT.

The coal-fired gas turbine with the PFBC is a proven concept in commercial operation. The hybrid steam plant with gas turbines is a viable system for cogeneration offering flexibility in balancing steam demand with fixed power loads in a way that that a steam turbine alone cannot. A cogeneration unit with gas turbine alone, or in a combined cycle format, would only be able to run heavy supplementary firing to achieve similar flexibility of heat output.

The other gas turbine link to coal is not direct but with a coal derived fuel. IGCC is not just a power plant, but a cluster of chemical processes with a combined cycle in the middle. The basic elements are an air separation plant which produces oxygen for the gasifier, the pressurized reactor vessel of the gasifier, and synthetic gas clean up, including sulphur and mercury recovery systems.

The air separation unit is independently supplied with air, although some gas turbines supply this air, about 15% of the mass flow, from the compressor, and the air separation plant sends the separated nitrogen back to the gas turbine combustor for flame dilution to suppress emissions. The steam turbine is larger than for a normal combined cycle because there is the cooling system of the gasifier itself as well as the syngas cooler placed between the gasifier output and and the syngas cleaning systems which remove sulphur and mercury. The gas is a mixture of carbon monoxide and hydrogen.

Coal gasification is a long established process in the coal producing countries of Europe and North America. Coal gas dates from the nineteenth century as a public energy supply in the major cities, initially for street lighting and later extending to cooking and space heating. The processes that created it were the building blocks of the chemical industry and of the coal gasifiers that have emerged a hundred years later.

IGCC as commonly understood refers to coal and some of the earliest examples were built in Europe and the United States. These were essentially prototype systems to test the technology and used relatively small gas turbines, although one of the latest, at Puertollano. Spain, has an F-class gas turbine: the SGT5 4000F rated at 245 MW, and 45% efficiency under ISO conditions.

All of the IGCC schemes are of a commercial size but the coal based schemes are few. Almost all were built as technology demonstrators. But in Europe the coal schemes were built as full size power plants, SUV Vresova in the Czech Republic, with the GE frame 9E in a 2+2+1 arrangement; Buggenum, Netherlands with the Shell gasifier and one Siemens V94.2 in a single shaft arrangement; and Puertollano, Spain, with a 1+1 multishaft block of Siemens SGT5 4000F. This last project has been using a mixed feedstock of locally mined coal and petroleum coke, and at 45%,

5.7 Vresova, Czech Republic: IGCC powers gasifier process to make clean transport fuels completely free of suphur. Is this the future for coal gasification? (Photo courtesy Siemens)

has the highest efficiency of the currently operating units.

In the United States, IGCC has been studied since the time of the first major oil crisis in 1973. Thirty-eight years later several projects have been built to demonstrate the technology, but the big factor is cost. IGCC is not specifically a coal gasifier; in fact the majority of schemes have been built at oil refineries gasifying the heavy, almost solid residues of the refining processes.

The other facet of IGCC is that it is potentially more than just a producer of electricity and process steam. With the many high sulphur coals which have been extensively tested in the existing plants, an essential part of the gas clean-up process is the production of elemental sulphur, which has a high commercial value and can be sold to the chemical industry.

Then there is the fuel flexibility of the gasifier which can be designed for firing coal, but also residual oil, biomass, pelleted refuse or combinations of those fuels. With the growing emphasis on recycling, the amount of domestic refuse is significantly reduced once the glass, metals, plastics, and clean paper and board have been taken out for recycling. What's left is more organic in nature and can be formed into pellets which can be blended with

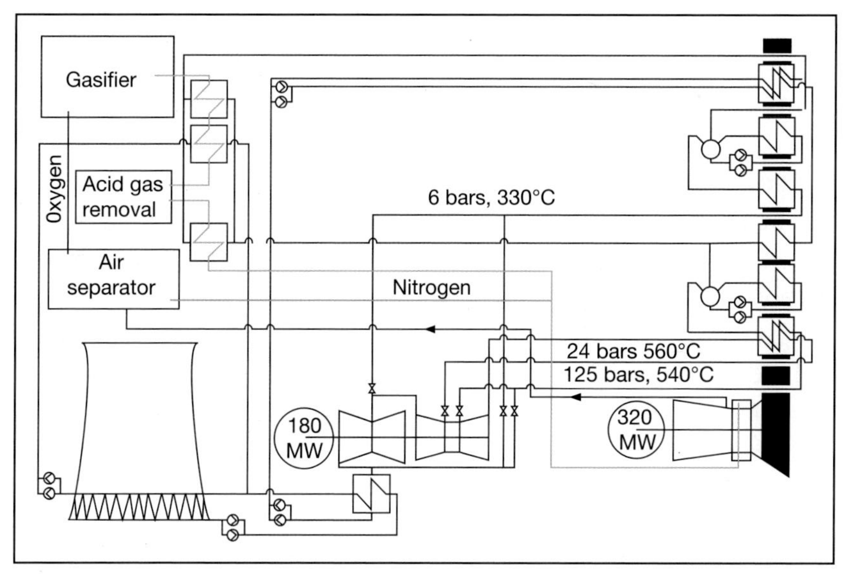

5.8 Schematic IGCC scheme in which gas turbine provides compressed air to the separator and receives back nitrogen for flame dilution and the Syngas cooler provides input to the steam cycle.

coal and burnt or gasified, thereby saving on transport to landfill.

Of the IGCC plants that have been built a number, particularly in Europe, are at oil refineries. The market for oil changes over time and processes must respond to need.

One of the biggest process changes in the last 20 years has been the switch to production of high-performance, unleaded gasoline, and particularly in Europe the production of increasing volumes of clean diesel fuel for pasenger cars. These processes result in different process endings; the near solid residual oils, asphalt and petroleum coke which are the feedstock for the industry's gasifiers.

An IGCC scheme at an oil refinery gasifies these process residues as fuel for a combined heat and power scheme which will supply power and process steam to the refinery processes. The majority of oil gasifiers are in Europe and are commercially operated by the refineries. They are among the largest units so far with two units of 500 MW in Italy and one of 400 MW in Japan.

In all there are currently more than 25 gas turbines in operation in IGCC schemes around the world with two million hours of combined operation between them. In the United States, the emphasis has been much more on coal gasification to produce fuel for a combined cycle power plant. These are all based on coal have been built as demonstrators to learn about the operation of the gasifier in the integrated system and of the performance of the combined cycle with the gasifier product gases.

The properties of IGCC are first that the synthetic fuel gas produces less carbon dioxide than even a natural gas fired combined cycle if carbon capture is considered. The performance of the pilot plants has suggested that the efficiency of the power plant without carbon capture will be between 40% and 45% (LHV) depending on the extent to which the syngas coolers are linked to the high pressure feedwater path of the heat recovery boiler.

In the eighteen years following the completion of the IGCC scheme at Buggenum, Netherlands in July 1994 twelve others were built, of which five in the United States, three in Italy, all at oil refineries, and one each in the Netherlands, Spain, Germany, and Singapore. The Spanish scheme at Puertollano, some 220 km south of Madrid, was completed in March 1998 has since been gasifying a mixture of locally mined coal and petroleum coke.

Of the five American schemes, Wabash River, and Polk County, use coal feedstock, and remain in operation. Around the world the coal based IGCC schemes, have converted different American and foreign coals with varying concentrations of sulphur, ash and moisture. The coal is converted by oxidation at high temperature and pressure which produces a gas which is a mixture of carbon monoxide and hydrogen, together with hydrogen sulphide and carbon dioxide. The molten slag can be cooled and sold as an inert construction aggregate. The gas cleaning process yields elemental sulphur and mercury which can be sold to the chemical industry.

The gas turbine burns a low-Btu gas with the flame dilution performed by nitrogen and steam. The nitrogen, in the case of an oxygen-blown gasifier, is returned from the air separation unit. Gas turbines used in the American schemes have so far been the GE Frame 7FA at Wabash River and Polk County; and the Frame 6B at El Dorado and Delaware City: Dow's IGCC scheme at Plaquemine, LA, used the Siemens W501D5.

When the first GE gasifier was installed at Cool Water, 150 km northeast of Los Angeles, in 1984 the cost was put at $2500/kW installed. But this was a prototype plant with the gas turbine of the day.

It was the first example of the Texaco oxygen blown gasifier and supplied a combined cycle consisting of a GE Frame 7EA with a 40 MW steam turbine on a separate shaft. The five American plants ranged from $1672/kW to over $2000/kW but in the ten years since these plants were installed costs have risen both for IGCC and conventional coal-fired steam plants, and the green dimension of clean coal has come to dominate.

Today IGCC projects under investigation for start up after 2015 are of the order of 600 to 800 MW. The combined cycle format would be a 2+2+1 arrangement of initially F-class gas turbines, with one gasifier train associated with each gas turbine.

The fact that IGCC has not progressed any faster is that there has been a considerable lack of enthusiasm until now among the coal burning power utilities. It is a power plant with a chemical process tied onto it. There are several gasifier designs each with different operating characteristics and maintenance requirements. What are the running costs of the gasifier and what is its availability compared with that of a typical combined cycle?

The other issue is the environmental pressures that have gone against coal in much of the developed world. Nearly half of the total installed coal-fired capacity is between thirty and forty years old and much of it running on subcritical steam cycles at efficiencies of around 30%. But it is natural gas fired-combined cycle that will initially replace much of this old coal-fired capacity. There is still a way to go before the first commercial IGCC plant will come into operation.

Meanwhile, what has changed is that the gasifier technology is being acquired by the power engineering companies. The aim is that the commercial IGCC power plant comes from a one stop shop with a specific gasifier as an add-on system optimized to a combined cycle.

In 2004 GE acquired the Texaco gasification technology from Chevron Texaco and entered an agreement with Bechtel with the aim of producing a standard integrated system linked to a combined cycle based on their Frame 7FB gas turbine. It was the same gasifier type as had been used in the first IGCC scheme at Coolwater, twenty years earlier, and is the most widely used gasifier system in the few schemes that have been built with six

5.9 Coolwater, CA: the first IGCC scheme to be built in the United States with a Texaco gasifier and supplying gas to a 110 MW combined cycle with a single GE MS7001E.

units, all at oil refineries. Four are in Italy and one, the largest, is in Japan which is linked to a 400 MW, single-shaft combined cycle block based on the Mitsubishi M701F.

In 2005 a deal was struck between General Electric, Bechtel, and American Electric Power (AEP) to perform front end engineering design for a planned 630 MW, IGCC plant on a site in Meigs County, Ohio. Since then, AEP have placed a second contract for a similar plant to be installed in West Virginia. Later Duke Energy also placed a front end design contract for a 600 MW plant to be installed at Evansport, Indiana. These initial design contracts were completed at the end of 2006 and first contracts could be awarded in 2007-08 for initial operation sometime after 2010.

Also at the end of 2005 GE announced a licensing deal with the China Petrochemical International Company (SINOPEC) to apply a gasifier to a chemical plant in Shandong Province. This was not an IGCC scheme but simply a gasifier for process application.

Development of clean coal technology has received funding from the US government as announced in the 2006 State of the Union message. The aim was to assist the development of the technology and in particular of carbon dioxide sequestration. IGCC produces

5.10 Lakeland, FL: Tampa Electric's Polk County IGCC plant is probably the cleanest coal-fired power station currently operating in the United States. (Photo courtesy of Tampa Electric)

a gas of which the principle component is hydrogen and carbon monoxide. For this reason the system must be dual fuel with either oil or natural gas for starting. Ultimately with CO_2 sequestration technology in place, the synthetic gas will almost all be hydrogen.

The commercial plant must therefore be designed to operate either on synthetic gas or natural gas. The steam turbine would be larger than normal for a natural gas fired plant because of the large high-pressure steam input available from the syngas coolers. It might only be a 2-pressure boiler since on natural gas firing it would run at part load to achieve the same steam conditions on the gas turbine exhaust energy.

The gasifier link to the combined cycle is to the high pressure steam path, and there are various ways in which it can be done. The most basic system is through the quench system of a slurry-fed gasifier which cools the ash leaving the vessel. Some gasifiers, notably GE Texaco, have a water cooled reaction vessel which can be cooled by high pressure feedwater. Another heat source is the free standing convective syngas cooler which has to reduce the temperature from about 750°C leaving the gasifier to about 50 °C going into the acid gas removal systems.

Gas cleaning removes acid gases - sulphur oxides - in a process which reduces them to elemental sulphur which can be sold to the chemical industry. Mercury can be removed by passing the gas over an activated carbon bed which can remove over 90% of any mercury present. Mercury removal is a statutory requirement of any IGCC plant installed in the United States. The cold, clean syngas can then be preheated on its way to the gas turbine using the same arrangement as with a natural gas fired combined cycle to raise its temperature to about 160°C.

Removal of carbon dioxide from the synthetic gas is possible and comparable applications in the chemical industries are running. An engineered system, installed in a commercial power plant is not yet available. The basis of it is a catalytic shift reaction with water which converts carbon monoxide to carbon dioxide and hydrogen. So the syngas of the future will be largely hydrogen, the combustion product of which is water.

GE and Bechtel are designing what in effect will be a standard design for an IGCC plant based on two oxygen-blown gasifiers supplying fuel to a combined cycle with two Frame 7FB gas turbines and a steam turbine. The normal S207FB combined cycle design has an output of 562.5 MW and an efficiency of 57.6%. As an IGCC system because the synthetic fuel is a low energy gas, with nitrogen for flame dilution, the mass flow is much higher so that each gas turbine produces about 40 MW more to achieve an output of 630 MW but at an efficiency of 39%. The GE design is limited to bituminous coal applications only.

Siemens is also developing a 600 MWe class IGCC plant based on its SGT6-5000F gas turbines for the North American market application for lower rank coals and lignite. Mitsubishi is developing a 450 MWe IGCC plant using its M501G gas turbine and an air-blown gasifier.

The low efficiency is calculated on the same basis as the net efficiency of any combined cycle: from fuel to saleable electricity. The heavy auxiliary loads of the gasifier and the gas cleaning system results in the low net efficiency, but it would seem to be clear from this that the initial design uses the quench mode of the gasifier with no cooling steam going to the combined cycle and only perhaps preheating of the fuel gas.

Despite the low efficiency there are advantages, none the less. It is a clean fuel which enters the gas turbine with no sulphur and no mercury. GE quote emissions of 39% less NOx, 75% less sulphur oxides, and 45% less particulates compared with the steam plant with its low NOx burners, and flue gas desulfurization and electrostatic precipitators. For the United States, at least IGCC represents the use of an indigenous fuel in a more environmentally friendly manner and guarantees continued employment in the mining industry. In fact the first commercial plants will be built in the traditional northern mining states.

At the end of 2010 IGCC was still uncompetitive against the alternative of supercritical pulverised coal when CO_2 capture is not considered. The net efficiency is marginally lower but there are many variables still to be evaluated before a standard repeatable format can be established. The oxygen blown entrained flow gasifier is a high temperature process which receives the coal in the form of a slurry or dry feed. The GE gasifier is a single stage slurry gasifier with a water walled vessel and is a radiant cooler which can be linked back to the steam cycle of the power plant. The Shell, Prenflo and the Mitsubishi air blown gasifiers are dry feed systems. The E-gas gasifier is a two-stage slurry gasifier.

The raw syngas gas leaves the gasifier at about 730°C and has to be cooled down to about 40°C for the acid gas removal process which cleans the gas of sulphur. One way is to use a convective cooler which also removes heat to the steam cycle. But systems used on some of the demonstration plants have had corrosion problems because they lie upstream of the acid gas removal unit and can also be contaminated with particulate matter which can lead to blockage of the cooling paths.

Of course, the more heat that can be removed to the steam cycle, the higher will be the net efficiency of generation. On the other hand so too will be the capital cost per kW.

In the United States which is likely to be the largest market in the foreseeable future, the most likely gas turbines to be used in the initial projects are the F-class machines which in their ISO power ratings on syngas are 232 MW both for the GE Frame 7FB and the Siemens SGT6-5000F; and would probably be used in a three-shaft combined cycle arrangement with one gasifier train serving

5.11 Puertollano, Spain: when completed in 1997 this was the most efficient IGCC plant at 45%. Prenflow gasifier with Siemens SGT5 4000 F in a 360 MW combined cycle. (Photo courtesy of Elcogas)

each gas turbine.

But GE have already done preliminary studies for their Frame 9H gas turbine which show that it could be possible to achieve 50 percent efficiency on a coal scheme. Certainly a more efficient combined cycle would contribute to this as also would a larger steam turbine linked into the syngas coolers. But there is a lot of development ahead, not least in the gas clean up system, before this can be realised. Also in a fully optimised system the use of a steam cooled gas turbine adds further questions to the linkage of the steam cycle to the syngas cooler.

A possible way to do this would be if the syngas coolers were two stage heat exchangers, which would keep the water cycle intact and separate from the actual syngas stream. Then there would be no problem of leakage between the syngas cooler and the steam cycle with its cooling streams going to the gas turbine.

Siemens also has significant experience of IGCC operation with coal, and with a number of designs of oxygen-blown gasifier. Their first IGCC project was Kellerman Lünen a 183 MW power plant for the German generator STEAG which was completed in 1972. This was based on a 74 MW gas turbine fired with syngas

5.12 Wabash River, IN, USA: one of only two IGCC still running in the United States. Since starting in 1997, it has gasified various high-sulphur coals and petroleum coke feedstock. (Photo courtesy of PSI Energy)

produced by a Lurgi gasifier with a heating value of 123 Btu/ft^3, and a composition of 12% hydrogen, 22% carbon monoxide with 55.6% nitrogen, plant efficiency was 35.1% about the same as the best of the conventional pulverised-coal plants at that time with a subcritical reheat steam cycle. Twenty years later Siemens also provided two W501D5 gas turbines for the DOW LGTI IGCC Project in Plaquemine, LA, that operated as an IGCC plant between 1987 and 1995.

Other European projects include the only two coal fuelled IGCC plants in Europe. The first of these was at Buggenum, Netherlands using the Shell gasifier, which went into operation in 1994. Based on the 150 MW Model V94.2, gas turbine it had a net efficiency of 43.1% and the gas had a heating value of 113 Btu/sft^3. The second, completed in 1997 at Puertollano, Spain, used the Krupp Thyssen Prenflow gasifier to power a 360 MW combined cycle with a V94.3 gas turbine. There was also an oil-refinery scheme similar to Buggenum at Palermo, Italy, but using a GE Texaco gasifier.

The 1997 take over of the Westinghouse non-nuclear power engineering business, brought them into contact with the American IGCC schemes. The company has worked with Fluor and Conoco

Phillips to supply two IGCC plants of 600 MW nominal output to sites in Illinois and Minnesota.

The Minnesota project, Mesaba, is at Iron Range, MN and received consent from the State Government in 2003. The steel industry in the region has declined with low world prices and more than 10,000 jobs have been lost in the last decade. The project owner and developer is Excelsior Energy and Siemens in partnership with Fluor and Conoco Phillips are the technology providers. Two E-gas gasifiers are planned each linked to an SGT6-5000F gas turbine in 603 MW combined cycle in 2+2+1 arrangement. But protests are mounting over the cost of the project and the cost of electricity that will result even without the addition of carbon sequestration and despite a $36 million Department of Energy Grant.

In May 2006, Siemens Power Generation bought the GSP gasification process, renamed the Siemens SFG gasifier, from the Swiss Sustec Group. It is an oxygen-blown gasifier which, for converting coal, is lined with water walled cooling screens, but has no refractory linings. The gasifier lining has lasted over 10 years in a unit at the SVZ Schwarze Pumpe Plant. Siemens is aiming for gasifier availability over 90%. The company claims that the cooling-screen design gives operational flexibility with quick start-up and shutdown, and that it can gasify coals with high ash-melting temperatures without affecting availability.

The Siemens SFG gasifier is one of several which were installed at Schwarze Pumpe, Germany, to gasify lignite and industrial wastes for the production of town gas in Cottbus and the surrounding area. Today, about 25% of the feedstock used at the facility is lignite and the rest various industrial and domestic wastes. After reunification, natural gas was brought in to replace town gas and in 1994 a 70 MW combined cycle with a Frame 6B gas turbine and a 26 MW condensing extraction steam turbine supplied by Hitachi, was coupled to the gasifiers. Then in 1999 a British Gas Lurgi gasifier was installed to replace the old units.

Having acquired the GSP technology Siemens announced that they would build a prototype at Spreetal, Germany. However, the focus will not be on IGCC, but on downstream synthetic gas conversion to produce clean, sulphur-free automotive fuels. Siemens has also taken an order from a Chinese customer to provide multiple gasifiers for a coal to chemicals project in China with start up of the first in 2010.

The other country which is making a concerted effort to develop the IGCC concept is Japan. As a net importer of coal, IGCC is important as a higher efficiency power generation system compared with a conventional steam plant with a supercritical steam cycle equipped with FGD.

The one operating IGCC plant in Japan is at Yokohama oil refinery, which uses asphalt as the feedstock for a Texaco gasifier which is integrated with a combined heat and power system supplying electricity to Tokyo Electric Power Company and process steam to the refinery.

The national development program is focused on the air-blown gasifier which is being developed specifically for power generation. It does not require an air separation plant and therefore has a lower auxiliary load and consequently a higher net efficiency from coal to electricity. The air-blown gasifier is a two-stage pressurized entrained flow design with a dry coal feed system. Since it is supplied with pulverised coal there is a low moisture latent heat loss compared to wet coal feeders based on coal slurry; and by using the same water-cooling wall structure in the gasifier as the boiler, it is possible to realize a highly reliable gasifier.

Mitsubishi has built a 250 MW IGCC demonstration plant at Iwaki, about 160 km north of Tokyo. The plant owners are Clean Coal Power Ltd, who are a consortium of nine Japanese utilities which are supplying 70% of the funding with the Ministry of Economy Trade and Industry supplying the other 30%.

This plant comprises a 1700 t/day gasifier and with the smaller M701DA gas turbine and was scheduled to start three years of operational testing in early 2007. Mitsubishi has sold several examples of the 701DA to the Asian steel industry where they are burning blast furnace gas. The air-blown gasifier also produces a low energy gas with an average calorific value of 1157 kCal/m^3. In fact there is little modification of the gas turbine required, apart from the compressor casing to be able to send about 17% of the air supply to the gasifier and fit the low energy gas burners.

The gasifier is a two stage process comprising a combustor and a reducing stage mounted above it. The combustor burns part of the coal to produce char and gases which carry it up to the reduction stage where more coal is added in a reducing atmosphere to produce syngas and more char which is collected in cyclones and recycled through the bed. The reaction is endothermic which reduces the final gas temperature.

With a smaller auxiliary load the air blown gasifier is should be more efficient in the larger commercial version based on the M701F gas turbine in a 3-shaft combined cycle arrangement at an efficiency of 45%, and 48% with the steam cooled M501G. But the aim is to develop the gas turbine to be able to send 100% of the air required to the gasifier and receive back 100% of the available inert gases for flame dilution.

The initial IGCC plants in commercial operation will have oxygen-blown gasifiers probably with an independent air separation unit sending nitrogen back for flame dilution.

One of the leading power plant design programs suggests that there are significant variations in output and efficiency depending on the cooling arrangements of the gasifier, the heat content of the fuel and it composition, and the amount of air sent to or nitrogen received from the air separation unit. The two gasifiers being studied for the American market are the GE Texaco, and the Conoco Phillips E-gas.

Both are two stage gasifiers which are fed with a pulverised coal/water slurry. In the first stage coal slurry is fed to the lower section with oxygen to maintain a temperature above the ash fusion point. This converts the coal to syngas which enters the second, upper stage, where more slurry is added. Char carried off with the syngas is trapped in a cyclone and recirculated to the bottom of the vessel.

The GE gasifier has a water-walled vessel which can be cooled by the high pressure feed water from the combined cycle. The syngas leaves at about 730°C and is further cooled to about 100°C for the gas cleaning systems. The E-gas gasifier has quench cooling and the gas leaves at about 1000°C The gas enters a large fire-tube type heat exchanger which is linked into the high-pressure boiler, and where the temperature is reduced to about 370°C. From both gasifiers the ash is a fine granular substance which can be used as a building aggregate.

The expectation is that these early IGCC plants will have an efficiency marginally higher than that of a conventional pulverised coal supercritical steam plant. But it really depends on how much heat is recovered from the syngas coolers into the high-pressure section of the steam cycle. The Texaco gasifier for instance can be supplied with just quench cooling, and some fuel gas preheating, which will give the lowest efficiency.

Alternatively the radiant cooler of the gasifier vessel jacket can be added to further raise the efficiency, and finally a convective cooler. All the cooling options could bring the net efficiency up to around 45% from coal to electricity sent out. If it means building a more complex system to achieve a higher efficiency, will the additional capital cost give a sufficient rate of return given the price of electricity that results, and can the reliability be sufficient that it does not compromise availability of the power plant?

With the current F-class air-cooled gas turbines and independent air separation units, a high pressure steam cycle and a high-quality bituminous coal the efficiency could be anywhere between 46 and 53% gross on the generator terminals; and 40 to 45% net depending on the extent of heat recovery from the gasifier vessel and the syngas coolers, and the amount of nitrogen returned from the air separation unit.

Gas turbines supplying air to the separation unit are being talked about and are a future development, but it would create a family of gas turbines designed specifically for IGCC applications and burning low energy gases. It does not of itself significantly raise the efficiency of the system and a far better way would be to apply higher powered gas turbines to lower costs.

The large cooling load represented by the gasifier and the coolers which drop the gas temperature by about 700°C between the gasifier exit and the acid gas removal unit, is a valuable source of energy which should not go to waste. The large volume of high-pressure water required for cooling duty suggests that a vertical two pressure heat recovery boiler with a large high-pressure flow, and with assisted circulation would not only avoid too large a volume of water but also reduce the cost of the boiler.

It may be more than fifteen years since the first IGCC plants were installed but the development of the system has not been so fast. Nevertheless with some of the current plans being offered with only 39% HHV efficiency, they are not using the cooling systems on the gas treatment side. Include these, and while they may increase the auxiliary load they will add to the high-pressure steam and therefore the steam turbine output.

In Europe the natural gas fired combined cycle with a single shaft block of an F-class gas turbine would have a gas turbine of about 280 MW and a steam turbine of about 130 MW which would have a net output of about 395 MW at 58% LHV efficiency. The same combined cycle in an IGCC scheme would have a higher gas turbine output because of the additional mass flow of the syngas and the flame diluting nitrogen returned with it, and a larger steam turbine because of the additional cooling load of the syngas in the clean up process. The total output would be 500 MW with 320 MW from the gas turbine and 180 from the steam turbine, with about 60 MW of the total output accounting for the large auxiliary load. But the net efficiency could be up to 46%, which, given that the fuel is coal, is a small gain in efficiency from fuel to electricity.

There is no standard gasifier although the gas turbine companies have taken up with specific designs for particular markets. There are six types which have been used in the current IGCC projects and two more under development. The GE Texaco, and ConocoPhillips (Destec) are slurry-fed whereas the Shell, British Gas Lurgi, Siemens SPG, Mitsubishi, and Prenflo are dry-fed with water used only to cool the slag in the bottom of the reactor. The AFB gasifier uses the principle of the fluidized bed boiler at high temperature and is a dry system with cyclones recirculating particulate matter entrained in the syngas. The Siemens SFG gasifier is also a dry fed system with a water wall screen in the reactor vessel. All the above are oxygen blown, except the Mitsubishi and AFB designs.

Dry coal feed is simpler, with less auxiliary load and so results in a higher efficiency. The Prenflo gasifier at Puertollano, Spain has run for seven years with a net efficiency from coal to electricity of 45%. Given that the whole purpose of the combined cycle is to generate electricity at a higher efficiency, to switch to a coal based scheme and not produce a worthwhile gain in efficiency, is a pointless exercise. Who would order an IGCC with a net efficiency of 45% when they can get the same from the proven technology of pulverised coal with a supercritical steam cycle?

There are still many issues to be resolved before it can be said that there is a competitive, IGCC plant in a standard configuration with a significant cost advantage over the traditional coal fired plants. Not least is the reliability of the heat exchangers with unclean gas passing through them and prone to corrosion and blockage. The challenge then is to get the reliability of the total gasifier system up to the same level as the combined cycle. If the major overhaul of a large F-class gas turbine takes place every four years, the gasifier and its auxiliaries must be able to run the same length of time between major overhauls.

First American design utility projects are intended for installation in the north eastern states: Ohio, Indiana, Pennsylvania and West Virginia. These are the traditional mining areas with high quality bituminous coals of heating values between 22000 and 35000 kJ/kg. Sulphur content is in the main less than 1%, but the notorious Illinois 6 has more than 4%.

There are also several projects planned for the western states that will use lower rank coals and lignite. A plant in Texas would use a locally-mined lignite with heating values of 8000 to 15,000 kJ/kg but low sulphur and high ash and moisture contents. Studies show that a lignite fuelled IGCC plant would have a higher output but a lower efficiency, all other things being equal.

While an export market for IGCC will take account of the coal sources available to the customer and the plant will be designed accordingly, the improvement of the efficiency and reliability of the system will drive development forward. One of the important issues is the reliability of heat transfer. There are water wall linings for some gasifier types, but the syngas coolers have a very clean fluid on one side with the high-pressure feedwater, and a very dirty gas on the other side. It requires stainless steel tubing to protect against corrosion but any solids in the gas may accumulate and block tubes or otherwise reduce cooling efficiency.

The other issue being considered by the gas turbine manufacturers is that of taking an air supply to the air separation unit from the gas turbine compressor. Already this is being considered for the air-blown gasifiers. About 15% of the total mass flow is taken from the compressor delivery and must be further compressed to reach the operating pressure of the gasifier. Taking air from the gas turbine compressor is a challenge due to compressor surge margins that must be controlled. However there is a smaller auxiliary load than is the case with an independent air separation unit and the gasifier reactions being exothermic a dry coal feed can be used, with a large potential for heat recovery to the steam cycle with a water-cooled reactor vessel and the syngas coolers.

The economy of clean coal technology is pointing to a higher price for electricity. Furthermore the existing plants are likely to close faster as legislation is introduced to control emissions as is being done in the United States. It is estimated that some 50 GW of plant will have to be closed.

Similarly the Large Combustion Plant Directive in Europe, which comes into force in 2015 will see only those coal-fired plants with Flue Gas Desulphurization and other control measures in place. So are we seeing the end of coal-fired energy in the industrial world?

Already large coal producing countries are looking at nuclear energy to fill the gap with its lack of gaseous emissions and a design life of sixty years. Poland is already looking at two nuclear reactors to start a programme. While in the United Kingdom there are plans to build up to ten reactors with the first in service by 2018. Meanwhile some combined cycles are planned to be in service before 2015 to run as base load plants until the new nuclear plants come into operation.

Coal gasification may continue to produce liquid transport fuels as several installations are in operation or planned. But maybe a small nuclear reactor will supply the energy for the process in the future.

6
Nuclear energy plans

By the end of 2010 there was definitely a nuclear renaissance gathering momentum. Several countries, including some with no nuclear plants in operation, had plans to build; and some, notably the United States, were inspecting existing units and granting life extensions to 60 years, which for most would mean a final shutdown after 2030. The industry could have seen some first orders which would have seen new plants based on new reactor designs coming into service in about 2017-18.

But on the afternoon of 11 March 2011 the most powerful earthquake ever recorded, 8.9 Richter, struck Japan about 130 km off the port of Sendai in northern Honshu. Eleven nuclear power reactors along the northeast coast automatically tripped and began to cool down to cold stand-by.

For these plants it was not unlike a shut down for annual maintenance, but not at Fukushima Daiichi which, with six BWR dating from the late 1960s had been one of the first nuclear plants to go into operation in the country. At the time of the earthquake, only three of the reactors were in operation. The others were down for maintenance, with their fuel transferred to storage ponds.

The severity of the earthquake produced a tsunami which headed towards the mainland and struck with a 20-metre high wave front which completely swamped the diesel auxiliary power plant and shut it down. It was about two months later that the consequences of the earthquake came to light. The island of Honshu has been physically moved about 4.5 m eastwards and 1.5 m downwards. Many towns on the northeast coast now see their streets flooded twice a day at high tide.

What subsequently happened was shown all around the world

on television. Reactions of some European governments were extreme and five countries are refusing to build nuclear plants and two are planning to shut all theirs down within the next twenty years if not sooner.

The Green anti-nuclear fanatics who are now represented in some governments are ecstatic. These decisions are essentially political but they also betray a lack of joined-up thinking in some governments. The countries are Austria, Denmark, Germany, Italy and Switzerland. The Italian decision is the result of a referendum in June 2011, just as the Austrian position was also the result of a referendum in 1978.

If the government is unpopular at the time of a referendum, the electorate will vote against it because they believe that it will hasten the time when they can vote it out of office. But it rarely happens like that unless the opposition leader is a popular and charismatic figure who can convincingly exploit the situation.

In Italy the reintroduction of nuclear power was one of four questions: the other three were on more political issues concerning an unpopular prime minister. In Austria the question was whether the nuclear power station at Zwentendorf, near Linz, should be brought into operation. They had an unpopular prime minister, for other reasons, who said he would resign if the vote went against him. The vote went against the prime minister who did not resign. Consequently no fuel was loaded into the plant which never went into operation, and was later dismantled and the equipment sold.

In Southern Germany earlier in 2011 in an election in the State of Baden Wurtemburg, where the Christian Democrats had always been the State Government for as long as anyone could remember, the party was resoundingly defeated. The result brought the Green Party into this state Government for the first time.

Parallel with this, in Germany, the reaction to events in Japan was to shut down six reactors which had been built before, or at the same time, as those at Fukushima Daiichi, i.e. up to the end of 1979. The German reactors are four PWR and two BWR totalling 6186 MW. All the reactors at Fukishima were BWR and of lower capacity.

Observers of German actions say that this was just a knee-jerk reaction. None of the shut down units are anywhere near the sea

TABLE 6.1: NUCLEAR POWER OUTPUT IN 2010

Country	*Operating reactors*		*Electricity production (TWh)*	
	Number	*Output (MW)*	*Nuclear*	*National total*
Argentina	3	1627	6.7	113.56
Armenia	1	376	2.3	5.84
Belgium	7	5943	45.7	89.43
Brazil	2	1896	13.9	448.39
Bulgaria	2	1906	14.2	42.90
Canada	18	12697	85.5	566.23
China	14	11271	70.1	3894.44
Czech Republic	6	3722	26.4	79.28
Finland	4	2741	28.4	100
France	58	63130	410.1	533.44
Germany*	17	20339	133.0	468.31
Hungary	4	1880	14.7	34.92
India	20	4385	20.5	706.89
Japan	51	44642	280.3	959.93
Korea	22	19785	141.9	440.68
Mexico	2	1600	5.6	155.55
Netherlands	1	485	3.4	100.29
Pakistan	3	725	2.6	96.29
Romania	2	1310	10.7	54.87
Russia	32	23084	159.4	932.16
Slovakia	4	1816	13.5	26.16
Slovenia	1	696	5.4	14.48
South Africa	2	1800	12.9	248.08
Spain	8	7448	59.3	295.02
Sweden	10	9399	55.7	146.19
Switzerland	5	3252	25.3	66.58
Taiwan	6	4927	16.80	207.38
Ukraine	15	13168	84.0	174.64
United Kingdom	18	10745	56.9	362.42
United States	104	101263	807.1	4117.86
Total	442	378058	26123	15307.57

and could never be subject to a tsunami. Moreover a technical inspection of the six units showed that there was nothing wrong that would prevent their being restarted. Earlier the Government had passed measures extending nuclear plant life to 2034. But electoral unpopularity has shifted closure back to 2022 the compromise closure date of the previous SDP/Green Government.

Denmark has no nuclear power plants and has been broadly anti-nuclear since the construction in Sweden of the Barseback reactors near Malmö, which are visible from tall buildings in

6.1 Brokdorf, Germany: sheep may safely graze beside this nuclear station near Hamburg, yet during the early years of construction it was the scene of violent protests by anti-nuclear factions. (Photo courtesy of E.ON AG)

Copenhagen and from coastal communities to the north. Sweden shut down both plants after 2000 when they depended on a Green minority party in government but in 2010 they announced that they would build new nuclear plants to replace the current ones at the end of their lives.

In May 2011, the Swiss government announced that it would shut down all its nuclear plants by 2034. Previously they had been considering two or three new plants of which the first would be at Beznau, to replace the two 265 MW PWR's which were the first to enter commercial operation in 1969 and 1971.

Four months later the Swiss Senate overturned the decision and entered a proviso that nuclear power should continue to avoid any serious power shortage, and its replacement with fossil-fired plant which would do nothing for the environment. New reactors would be built but only the currently available Generation III designs. That means the Areva EPR, the Westinghouse AP 1000, or the GE ABWR. As the Senate came to its decision Atel, who hold 40% of Gosgen Daniken, filed a plan for Niederamt, which would be built on a neighbouring site. This would be either EPR or AP1000.

Earlier, a consortium of Nuclear Utilities had filed application

6.2 Biblis, Germany: these two 1200 MW reactors completed in 1977 and shut down post Fukushima may never operate again. In any other country they might still be operating with a life extension to 60 years. (Photo courtesy of RWE)

for EPR to be built at two of the other nuclear sites: Beznau and Muhleberg, and it is these two plants which have been stopped by the Lower house. Currently nuclear power meets 40% of total electricity demand with the rest hydro and imports from Germany and France. A general election in October 2011 is expected to resolve the issue and if not it may be put to a public referendum.

Events at Fukushima therefore, have had a profound political effect on the nuclear industry of Europe. No other countries have announced abandonment of nuclear power and so the probability is that all the other nuclear operating countries will continue with rheir existing plants, and build new ones according to need.

Then, in the afternoon of July 2, 2011 a transformer exploded at the Tricastin power plant in France causing oil to burn and raise a huge black cloud of smoke. The transformer was not associated with the nuclear reactors but it had nevertheless happened on a nuclear site. So it brought the anti-nuclear fanatics out in force.

It is only in Europe that outright public hostility to nuclear power exists. If Germany closes everything in 2022 and there is no EPR built at Beznau by then, and with the rest of Switzerland shutting down in 2034, there will soon after be a solid group of countries extending from Denmark down to Sicily with no nuclear energy production and a large block of base loaded generating capacity taken out of service for entirely political reasons, and dependant on imports from France, Belgium, and the Czech Republic, and upon unreliable renewables to reduce their emissions.

Japan is rather a special case in more ways than one. Anybody visiting the country is sure to experience an earthquake. There are no windows open in your hotel room but suddenly the curtains sway and that is the sign of a very mild earthquake shaking the building, of which there are many every year.

The 2011 earthquake was stronger but did not kill anybody; it was the following tsunami which did all the damage as it swept ashore. Having seen on television the hydrogen explosions in the reactor buildings at Fukushima Daiichi and the spraying of water into the damaged spent fuel pond, coupled with the comments as to what might happen if they couldn't control it, it was enough to have some countries looking at their nuclear plans again and hoping for the inevitable shut down.

In fact Japan is one of the few countries which has most of its nuclear power plants on the coast. This is not true of France, Germany, Belgium, Canada, the United States and Russia, to mention a few. In Belgium, the four plants at Doel are on the Schelde Estuary, and in Germany Brokdorf is on the estuary of the Elbe, northwest of Hamburg, and is their only coastal nuclear plant. In Canada, their only coastal power plant is at Point Lepreaux, New Brunswick on the west shore of the Bay of Fundy and sheltered by the mainland of Nova Scotia from the full force of the Atlantic weather.

Other than Japan, which countries could experience a tsunami at their coastal nuclear power sites? In fact a map of the nuclear plant locations in the world shows that there are only four countries with nuclear plants on exposed sea coasts: the United States, Brazil, South Africa and Japan. Sites in the rest of the world although on the coast are sheltered by neighbouring countries. Not even the United States, which has two plants on the Pacific coast, noticed the tsunami which struck Japan because the wave crossing the Pacific would have been damped down over the long distance.

But this was no ordinary nuclear accident, it was a natural

event of unprecedented severity which prevented the orderly shut down of the operating reactors by flooding the back-up diesel plant. Then again, it was only Fukushima Daiichi where they had problems shutting down, and in the light of what has since been discovered was this a part of the coast which has sunk as a result of the earthquake? The other aspect of the plant is its age and that the reactors were all early model BWR's over thirty years old.

A number of plants are installed in seismically active areas and have survived earthquakes without incident. But then we look at the Great Kanto earthquake of 1923, which killed 140,000 people in and around Tokyo. Since then civil engineering has advanced in its understanding of building design to resist earthquakes and in many countries it is the historic buildings, churches, palaces, temples and the like which predate 1900 that often sustain earthquake damage.

So the slowing of European nuclear plans is not because they are subject to the risk of tsunami and not because they are planning to build new reactors! It is rather panicking governments seeking to assure their electorate that the nuclear plants proposed are not safe and the existing units are unable to withstand what in Europe at least are impossible natural disturbances. Also few governments contain engineers who might understand nuclear technology and be able to explain it to the people.

With 104 operating reactors and two more under construction the United States has more nuclear power plants than any other country which accounted for 20.15% of electricity demand in 2010. But it is at a loss as to what to do with its nuclear waste, including military and medical waste, which is at present stored at 133 sites dotted around the country including the nuclear power stations.

This has been gradually building up since 1989 when President Jimmy Carter, who had done military service as an officer on a nuclear submarine, and could have been expected to have understood something of the technology, cancelled a planned reprocessing plant under construction at Barnwell, NC, and ordered all nuclear operators to use a once through fuel cycle.

Although Carter himself, could have been anti-nuclear, the Democratic party which he led certainly had many anti-nuclear members. Like most parties of the political left at the time who

6.3 Watts Bar, TN: construction of two Westinghouse 1100 MW PWR was stopped in 1988, but Unit 1 went into service in1996 and Unit 2 is due to follow in 2012. (Photo courtesy of Tennessee Valley Authority)

advocated nuclear disarmament, it had been infiltrated by the Green movement, with their enthusiasm for renewable energy and fear of plutonium. They didn't understand that plutonium which was contained in the spent fuel and which, 44 years earlier, had been used in the Nagasaki bomb, was itself a valuable fuel material which could be used as mixed oxide fuel in the existing reactors, as has been done many times in Europe and the Far East.

Meanwhile a repository for spent fuel and nuclear waste has been built at Yucca Mountain, NV, and is ready to receive waste. But Nevada has a Democratic State Government who do not want nuclear waste to be left there. One of the imagined problems is that in transporting the waste across the country by rail it would come to a level crossing, and collide with a vehicle and become derailed, and the material for a dirty bomb would be stolen from the wreckage by some home grown terrorists, who had a grudge against American society, and would detonate it in a city instead of shooting a few people in a supermarket, or some other public place.

The current Democratic President, Barack Obama, appointed a pro-nuclear Energy Secretary; authorised the first Construction

6.4 Bellefonte, AL: Construction of this plant was also halted in 1988 due to slow growth in electricity demand. A decision was taken in 2011 to complete Unit 1. (Photo courtesy of Tennessee Valley Authority)

and Operation Permits, and sought funds for loan guarantees for new plants in the 2011 budget, has nevertheless sided with his party members in Nevada on the use of Yucca Mountain as a nuclear waste repository.

So there is nowhere that American utilities, hospitals and armed forces can dump their nuclear waste. A blatant political decision has blocked Yucca Mountain and an earlier decision prevents the nuclear operators from sending spent fuel to Europe for reprocessing. Were this to be done a large inventory of material could have been returned as mixed oxide fuel for their existing reactors. Worries about transport of nuclear waste are more imagined than real. Thirty years ago this problem was addressed in the UK where the majority of the nuclear plants were in the south of England and the Sellafield reprocessing plant was in the north. A special heavy steel container was produced to carry the spent fuel assemblies by rail and a series of tests were performed in which a diesel engine pulling a truck carrying a container was deliberately crashed on a disused branch line to see what would happen. The container came out of the derailment unscathed and it was only the locomotive that was seriously damaged.

TABLE 6.2: STATUS OF US NUCLEAR PROJECTS, 2012

Constructor	*Site*	*Output MW*	*Reactor*	*Service Date*
Under Construction				
TVA	Watts Bar 2	1218	PWR	2013
TVA	Bellefonte 1	1200	PWR	2015
Southern Nuclear	Vogtle 3 & 4	2400	AP1000 x 2	2017
SCE&G	Summer	2400	AP1000 x 2	2018
Loan Guarantee short list				
NRG Energy	South Texas	2712	ABWR x 2	2017
UniStar Nuclear	Calvert Cliffs	1710	US EPR	2017
Progress Energy	Levy County	2400	AP1000 x 2	2022

New American nuclear projects are shown in Table 6.2. Of the two under construction Watts Bar is a TVA project stopped in the 1980s when electricity demand fell below estimates. It has two 1200 MW Westinghouse PWR of which the first was completed in 1996. Construction of Unit 2 was halted in 1988 when 80% complete. In 2007 a TVA Board decision was taken to complete Unit 2 which is expected to enter service in 2012.

TVA also halted construction on their Bellefonte power station and later suggested adding two Westinghouse AP1000 PWR's to the site. The decision has been taken for the moment to complete construction of Unit 1 which would take care of the immediate load growth in the area. TVA will sell Watts Bar 1 to fund the job, but will stay on site as the plant operator.

Southern Nuclear's Vogtle station south of Augusta, GA, has two 1200 MW Westinghouse PWR which were completed in 1987 and 1988. Both have recently received life extensions to 60 years. The new project will install a further two reactors of the AP1000 design, four examples of which are already under construction in China. The Vogtle reactors will be the first of the type in service in the United States in 2016 and 2017 respectively.

Georgia Power the largest shareholder in the plant and the ultimate operator, issued a contract to Westinghouse and Shaw Group in April 2008 and in August 2009 the Nuclear Regulatory Commission issued the Early Site Permit and a Limited Work Authorization. In February 2011 a $14 billion Loan Guarantee was awarded as the last step leading to a Construction and Operating

Licence, which was awarded to Southern Nuclear on February 10, 2012. The $14 billion project is scheduled for completion in 2018.

The other four plants in the table are in line for the next four loan guarantees. The South Texas Project will be the first to use GE's Advanced Boiling Water Reactor, of which four examples are operating at three sites in Japan. Calvert Cliffs will have the first example of the 60 Hz version of the European PWR, of which there are two under construction in China and one each in Finland and France. The Finnish plant will be the first to be completed and is planned to be in operation in the winter of 2012-13.

The nuclear revival is beginning to gather momentum with more companies appearing as vendors with new reactor designs. The Middle East is expected to expand rapidly, with Egypt, Turkey, the United Arab Emirates and later Saudi Arabia, all expected to place orders in the next ten years.

The new global vendors are from Korea, Japan and Russia all with offers of PWR. The Korean Design is based on the original Combustion Engineering System 80 design. This was a 2-loop, PWR design, which was installed as units 5 and 6 at Yonggwang, and have been in operation for ten years, the last going into operation in December 2002. These are designated OPR-1000.

The uprated version, known as APR1400 is initially rated at 1350 MW. Four units are under construction for two Korean sites with the first, Shin Kori 3, planned for operation in September 2013. It is this design which was subject to a contract with the UAE, signed in Abu Dhabi by the heads of state at the end of 2009. For Korea the contract is worth $40 billion, of which $20 billion is for operation and maintenance over the first 20 years.

Electricity demand in the UAE was 15,000 MW in 2010 and much of it supplied by coastal combined cycle plants which have back-pressure steam turbines driving multi-stage flash distillation units to produce the public water supply. But with population and industry growth, electricity demand is expected to double across the Emirates and reach 40,000 MW by 2020. They want nuclear power to help meet this demand in an environmentally friendly manner. The four reactors will be sited on the Gulf coast at Bakra a sparsely populated area near the Saudi Arabian border 150 km south of Dubai and will have a combined electrical output of 5600 MW. Limited site preparation work was already in progress at the end of 2011.

To those Green fanatics that still argue that nuclear power is unsafe and must be ended and its placc taken by renewables and conservation, they don't understand what is happening in the UAE. What else is production of the public water supply from the sea than a prime example of conservation of energy? However with a growing population and industrial development, demand for electricity is outstripping the demand for water so what else can they do?

Work on the first unit starts in 2012 for completion in 2017. All four reactors should have been completed and in operation by 2020. In Korea the last of the four similar reactors, Shin Ulchin 4, is due to go into operation in 2016, one year before the first plant in the UAE. Any issue which might arise in the operation of the four Korean units can therefore be read across to those under construction so that any modifications can be incorporated as necessary.

Japan, with 34% of its electricity coming from nuclear plants is now tied to the American nuclear industry. Westinghouse is now owned by Toshiba, while GE has strong links to both Toshiba and Hitachi, former licensees for BWR, in the development and construction of the ABWR's in Japan. Meanwhile, Mitsubishi Heavy Industries (MHI) are bidding for two projects in the United States with their Americanized 1700 MW USPWR. Dominion Energy has chosen it for North Anna 3, in Virginia, while Luminant Energy want two for their Commanche Peak, TX, site.

The immediate problem for Japan is to bring into operation the nuclear plants which were shut down in March. The tsunami destroyed much of the grid system where it struck and this is not the only reconstruction which must be performed.

For Russia, the Chernobyl accident and its consequences led to cooperation between the Russian and European nuclear industries to bring Russian PWR and RBMK designs up to western safety standards. As a result the Russian nuclear industry is bidding against western companies in the World market, and in particular to new markets where nuclear energy is being considered for the

TABLE 6.3: RUSSIAN NUCLEAR PLAN TO 2020

Site	*Reactor*	*Output MW*	*Service date*
Kalinin-4	VVER-1200	950	2011
Beloyarsk-4	FNR BN-800	750	2012
Novovoronezh II-1	AES-2006	1085	2012
Sosnovy Bor 2-1	VVER-1200	1100	2012
Sosnovy Bor 2-2	VVER-1200	1100	2013
Volgodonsk-3	VVER-1200	950	2014
Baltic1	AES-2006	1200	2015
Seversk 1	VVER 1200	1200	2016
Sosnovy Bor 2-3	VVER 1200	1200	2016
Nizhegorod 1	VVER 1200	1200	2019
Seversk 2	VVER 1200	1200	2018
Tver 1	VVER 1200	1200	2017
Nizhegorod 2	VVER 1200	1200	2022
Tver 2	VVER 1200	1200	2017
Baltic 2	AES 2006	1200	2017
Sosnovy Bor 2-4	VVER 1200	1200	2019
Tsentral 1	VVER-1200	1200	2018
Tsentral 2	VVER-1200	1200	2019
Beloyarsk 5	BREST	300	2020
Dimitrovgrad	SVBR-100	100	2020

first time, as well as to their customers in the former Communist countries of eastern Europe which have now joined the European Union.

The first of the new generation of Russian reactors is under construction as the first of ultimately four units at Novovoronezh. Two more are under construction in the Kaliningrad enclave between Poland and Lithuania. Known as the Baltic plant its construction is critical in the region, first to improve security of electricity supply in the enclave, but also to export to neighbouring countries.

Poland with nearly all its capacity coal-fired must shut down its oldest coal-fired plants at the end of 2015 under the terms of the European Union's Large Combustion Plant Directive. The RBMK plant at Ignalina, in Lithuania has been shut down since the end of 2009, and plans to build a new plant on the neighbouring site of Visagenas have reached a decision in favour of General Electric's

ABWR. The Baltic power station is therefore designed to supply both the Kaliningrad enclave, replacing gas-fired capacity, and also to export electricity to Poland and the Baltic states. The project has been brought forward so that the first unit will be in operation in 2015. The plant has two reactors of type AES-2006 which are nominally rated 1150 MW. They are being built on a coastal site at Nemen, 20 km below the Lithuanian border.

The AES 2006 reactor is their latest 1000 MW design for which Alstom have contracted to supply their Arabelle turbine used on the French nuclear plants of similar capacity. In the spring of 2011 eight reactors were under construction: seven PWR and a third fast neutron reactor at Beloyarsk. These are the first units of a programme which aims to increase the nuclear share of electricity supply from 16% in 2007 to at least 25% by 2030.

Russia is starting to move out into international markets. They have already two 1000 MW units operating in China at Tianwan, since 2004 and 2005, respectively and in 2010 a contract was signed for two more reactors on the site, which also has room for another two. All six reactors should be operating by 2020. Two more 1000 MW reactors are nearing completion in India at Kulankulam, in Tamil Nadu state. Ultimately there could be eight reactors on this site.

Russia will build the first nuclear plant in Turkey, which will be at Akkuyu on the eastern Mediterranean coast near Mersin. Preliminary site works started in March 2011 and when completed there will be 4x1000 MW reactors on the site. Bangladesh is short of capacity by about 2000 MW and a contract has been signed with Rosatom for two 1000 MW reactors to be built at Rooppur, in the northwest of the country.

Russia is the only country at present to be operating a Fast Neutron Reactor (FNR). The 600 MW BN 600 went into operation in 1980 and in 2007 it was shut down for a comprehensive turbine and instrumentation upgrade and was then given a life extension of fifteen years, which anticipates final shut down in 2025.

Russia has also sold two BN800 reactors to China who want to use them to consume plutonium and higher actinides in the spent fuel from their existing reactors. The BN800 under construction at Beloyarsk will be a fast neutron reactor and is scheduled to come into operation in 2012. The Fast Neutron Reactor is in fact the commonly understood FBR, liquid sodium-cooled reactor with a plutonium core and a blanket of uranium or spent fuel. Fast neutrons from the plutonium core strike atoms in the blanket which include uranium and long-life actinides which can be fissioned to produce more energy.

With the complete fuel cycle and the reactors that can exploit it, Russia has a far more practical stance on nuclear energy than the other countries that initially developed it. France and the UK have also developed the fuel cycle, but the UK did so against a growing chorus of anti-nuclear protest as it built THORP the Thermal Oxide Reprocessing Plant at Sellafield, which went into operation in 1994.

Eleven years later THORP was shut down in 2005 after a leak of highly-radioactive liquid containing uranium, and plutonium in concentrated nitric acid. In 2004 there was noticed a discrepancy in records of plutonium passing through the plant. There appeared to be 29.6 kg of plutonium missing which was found to be in a sump of the clarification cell. It had always operated at below full capacity, and British Nuclear Fuels Ltd (BNFL) had lost £1.2 billion in reprocessing some 7000 t of spent nuclear fuel.

All this happened during the tenure of office of a Government which appeared to be anti-nuclear. It was decided that BNFL would become the Nuclear Decommissioning Authority and would stop reprocessing at THORP in 2010. The original reprocessing plant for spent Magnox fuel will continue in operation until the last two Magnox power stations, Oldbury north of Bristol and Wylfa on Anglesey, are shut down.

Following Kyoto the European countries sought to reduce greenhouse gas emissions by 20% of 1990 levels at the end of 2010. Despite a large programme of off-shore wind farms, it was explained to Prime Minister Tony Blair, that nuclear power with no emissions was the way to go. But these discussions paralleled the problems at Sellafield, so that when Gordon Brown succeeded him in 2007, it was decided that future nuclear power stations would run on a once-through fuel cycle.

The present coalition Government has extended the use of THORP to 2025 and has narrowed its reactor choice to the Areva

6.5 Sellafield, UK: after reprocessing nuclear waste is returned to the country of origin. Three nuclear containers carrying 76 canisters of waste immobilized in borosilicate glass blocks leave for Japan. (Photo courtesy of NDA)

EPR, and the Westinghouse AP1000, both PWR's which will be in operation elsewhere before the first enters service in the UK in about 2018. Initially ten reactors will be built on existing sites by three companies.

British Energy, a division of EDF, has plans for two 1600 MW EPR each at Hinkley Point and Sizewell. Horizon Energy, a joint venture of E.ON and RWE, plan either two EPR or three AP1000 at Oldbury, north of Bristol, and either three EPR or four AP1000 at Wylfa, on Anglesey. NuGeneration are a partnership of Iberdrola, GDF Suez, and Scottish and Southern are planning three reactors on the new Cumbrian sites.

Since Iberdrola now own Scottish Power, NuGeneration in fact is the Scottish Utilities in partnership with GDF Suez. Therefore, given that the Scottish National Party, which now runs the Scottish Regional Government, is basically anti-nuclear the likelihood is that much of the output of the Cumbrian plants will be sent north over the border.

But the saddest case of all is that of the United States which has virtually abandoned the nuclear fuel cycle. But in the beginning things were so different. West Valley, NY, about 60 km south of

6.6 West Valley, NY: the only operational reprocessing plant ever built in the United States which ran from 1966 to 1972. Other reprocessing plants have followed but have not been allowed to start up. (Photo courtesy of NYSERDA)

Buffalo was the site of the only reprocessing plant ever built and operated in the United States. It was built on 1352 hA of State owned land, and went into operation in 1966, eight years after the first nuclear power station went into operation at Shippingport, PA. In the next six years the operator, Nuclear Fuel Services Inc., reprocessed some 640 tons of spent fuel from Federal Government defence reactors and commercial nuclear power reactors and created two repositories for low- and high-level radioactive waste.

It was shut down in 1972 in order to increase reprocessing capacity and reduce radioactive effuents and operator radiation doses. The shut down occurred as new regulations were introduced covering earthquake and tornado protection and new requirements for waste management and Nuclear Fuel Services concluded that it would not be economically viable to conform to these requirements and continue reprocessing in the modified plant and ceded the site to New York State.

In 1975, New York State Energy Research and Development Association (NYSERDA) was created and in 1980 Congress passed the West Valley Demonstration Project Act. Under this act

NYSERDA would build a waste encapsulation facility on a 71hA site in which to provide the following:
* Demonstrate solidification of high level radioactive waste in the underground waste tanks.
* Transport the solidified waste to a federal repository for final disposal.
* Dispose of low-level and transuranic waste.
* Decontaminate and decommission the solidification plant.
When the plant was shut down in 1972 there were 750 spent fuel assemblies awaiting reprocessing and almost 1820 m^3 of highly active liquid waste, held in underground stainless-steel tanks.

By 2002, 98% of the waste had been solidified and had been packaged in 19,000 drums of cemented low-level waste, and which were sent to the Beatty, NV test site, and 275 stainless steel containers containing borosilicate glass logs of high-active waste which are stored in a shielded cell of the former reprocessing plant, until such time as it is deemed politically correct to send it to Yucca Mountain, but the energy content of the spent fuel is lost forever.

Attempts to commission new reprocessing plants were thwarted during the Carter Presidency at the end of the 1970s. In 1967 GE was authorised to build a 300 t/year plant, presumably to reprocess BWR fuel. It was completed in 1972, but never put into operation. GE now uses it as a spent fuel store.

A much larger reprocessing plant for 1500 t/year started construction in 1970 but was cancelled in 1979 because President Carter had reached a policy decision to abandon reprocessing and force nuclear operators into a once through fuel cycle.

The 1979 election brought a change of Government with the Republicans back under Ronald Reagan one of whose first energy acts was to lift the Carter ban on reprocessing. But it was not until 1992 that his successor, George Bush Sr., was confronted with a utility who wanted to send spent fuel to France for reprocessing, and refused to allow Long Island Power to make a contract with COGEMA.

So the situation today is that reprocessing is considered to be too expensive to pursue and that no commercial plant should be built until the best technology is proven by research. But there is a further issue which is peculiar to the United States. Energy policy is dictated by the President and not by the Department of Energy placing a motion before Congress which would then be debated in both houses and result in a specific energy act.

But the big international environmental movements began in North America at a time when nuclear energy was beginning. The fact that reprocessing produced plutonium was anathema to them and their growth of infuence in political circles created a national fear of plutonium which didn't recognise it as a fuel material in its own right.

Whether this will ever be resolved is open to question because every nation that has detonated a nuclear weapon, including India and Pakistan, must have some reprocessing technology of their own. Yet in Western Europe, Russia, India and the Asian Pacific countries reprocessing is being carried out for their nuclear plant operators who get more energy out of their uranium resources.

The three remaining American Reactor firms, Westinghouse (now owned by Toshiba), GE-Hitachi Nuclear, and Babcock & Wilcox, have continued with the development of new reactors and would appear to have taken the view that if they cannot sell in America they can certainly sell abroad, which certainly GE and Westinghouse have done in the Far East and linked up with their Japanese nuclear licensees. Mitsubishi remains independent in Japan and similarly, Babcock & Wilcox in the United States.

So in a way the tables have been turned as they have tied up with companies whose customers have experience of reprocessing their reactor fuel in Europe, and now in Japan. So if you buy an American reactor you can get your spent fuel reprocessed, provided you are not an American operator.

So what of the new designs? There are five PWR, from Areva, Rosatom, Westinghouse, and Mitsubishi, a BWR from GE Hitachi and the Advanced CANDU Reactor from Atomic Energy of Canada.

The PWR designs are already in production with Westinghouse having sold four in China and six to three sites in the United States. Two sites, Summer in South Carolina, and Progress Energy at Levy County, FL await Loan Guarantees, along with Areva's Calvert Cliffs, MD, project for Unistar Nuclear.

Areva has three projects of which the first is in Finland at

6.7 Tianwan, China: first two Russian designed 1000 MW reactors are nearing completion for service in 2013. Ultimately four more units will be added to the site. (Photo courtesy of CNNC)

Olkiluoto which is expected to be in service at the end of 2012. One is now under construction in France at Flamanville, on the Cherbourg Peninsula. A third unit is also being planned for Penly, where there are currently two 1300 MW PWR which have been in operation since 1990 and 1992 respectively.

Rosatom, also has examples of their AES 2006 reactor under construction in Russia. The first two are in the second plant at Novovoronezh now under construction. These are a standardized design of their VVER I000 PWR for the international market. The first unit is due to go into service in 2012, with the second following a year later. Construction of two units has started for the Baltic station near Kaliningrad, of which the first has been brought forward for completion in 2015 with the second in operation in 2017.

Westinghouse have four of their APR 1000 under construction at two sites in China, at Haiyang and Sanmen. The two projects are running in parallel with the first unit on each site in operation in 2013 and the second following a year later. Meanwhile, their first American project, Vogtle 3 and 4, for Southern Nuclear has started preliminary site work. On 16 February, 2011 loan guarantees for

6.8 Sanmen, China: the pressure vessel of the first reactor is lowered through the top of the containment building. Reactor is Westinghouse AP1000 of which six units are planned for the site. (Photo courtesy of Westinghouse Nuclear)

$8.3 billion were awarded to Georgia Power who hold 45% of the capital in the site and a similar entitlement to power from the existing units, and will operate the new plant.

GE-Hitachi have four examples of the ABWR in operation in Japan and two more nearing completion in Taiwan. This was the first of the new reactor designs to get NRC certification. The first two units went to Tokyo Electric Power Company's Kashiwazaki Kariwa station as units 6 and 7 which were commissioned in 1996 and 1997, respectively and operated until the station was shut down in 2006 by a powerful earthquake with its epicentre only a few kilometres from the site. Three years later, in May 2009, the Prefectural Authorities authorised Unit 7 to be started up as the first of the station's seven BWR's.

Elsewhere, Hamaoka 5 was the next Japanese ABWR to enter service in 2004, followed by Shika 2 in March 2006. Meanwhile, on Taiwan two ABWR are planned for operation in 2011 and 2012 at Lungmen on the north coast of Taiwan near Taipei.

There were two independent design routes to BWR: GE in the United States and ASEA Atom in Sweden. Other BWR Vendors were licensees of GE in Europe and Japan. All have contributed to the current ABWR, and particularly from the Swedish designed reactors of which there are ten operating in Sweden and two in Finland. All twelve reactors have received upgrades and the two Finnish plants have been granted life extension to 60 years.

Three specific features of ABWR are the circulation pumps and the reactor controls. The circulation pumps are now mounted in the bottom of the pressure vessel with only the pump motors outside the pressure vessel. This concept was first used on the later units of the ASEA Atom BWR in Sweden and Finland in the 1970s. In the ABWR there are ten units which replace the two large external jet pumps and supporting pipe work used in all the earlier GE BWR's.

To the control rod adjustment capabilities the electro-hydraulic Fine Motion Control Rod Drive has been added, which allow for fine position adjustment, while not losing the reliability or redundancy of traditional hydraulic systems which are designed to accomplish rapid shutdown in 2.80 seconds from receipt of an initiating signal.

Fully digital reactor controls (with digital and manual backup systems) allow the control room to easily and rapidly control plant operations and processes. In particular, the reactor can go critical and ascend to power using automatic systems only and similarly shut down using automatic systems only. Of course, human operators remain essential to reactor control, but much of the work of bringing the reactor to power and descending from power can be automated at operator discretion.

GE-Hitachi with the long term operation of their new design and in Japan, a country susceptible to frequent earthquakes, have a pool of operating experience which their competitors do not yet have. However the ESBWR (Economically Simplified Boiling Water Reactor) will be the last to get final design certification.

The reactor designers have gone for simplification to reduce cost and construction time. ESBWR for instance is a natural circulation design with no circulation pumps as such, and a passive saftey system which is gravity driven.

At present the only Westinghouse reactor in the UK is Sizewell B on the Essex coast which has been in service since 1995. At present it has a 40-year operating license which could be extended to 60 years as have many similar reactors in the United States.

The first AP 1000 may be not at Sizewell, but on Anglesey at Wylfa where the last of the Magnox stations will shut down in 2012. RWE, in partnership with E.ON, plan a reactor to replace the present unit, and similarly at the other operating Magnox site, Oldbury, north of Bristol. But British orders will not happen until the Generic Reactor Design evaluation is completed by the British Government, expected to be in June 2011. The same applies to the Areva EPR and these two reactor designs are the only ones being considered for the British nuclear power programme.

Westinghouse is already gaining orders in existing markets. China Power Investment Corporation is planning to buy up to twelve more AP 1000. The first four will be at their Pengze site in Jiangxi Province. Four more reactors may be added to the existing Haiyang site in Shangdong, where there are two AP 1000 under construction. It was planned to have six reactors on the site.

Areva's 1600 MW EPR is currently the largest of the PWR's under construction. The first two plants in Finland and France have had problems in construction which have delayed the schedules so that Olkiluoto 3 will not now go into operation until the winter or 2012-13. However, it must be recognised that these were the first nuclear plants to be built in Finland and France for more than ten years. So that there has been a lot of training in documentation and construction as the industry has started to rebuild.

Areva have plans for initially four reactors in the UK, two each at the existing sites of Hinkley Point, Somerset, and at Sizewell. They are also facing competition from Westinghouse with the smaller AP 1000 reactor.

But their biggest project could be in India a country which in 1974 exploded a nuclear device in the Rajasthan desert. Canada which in 1973 had completed the first of two 220 MW CANDU reactors immediately withdrew all technical cooperation and from then, with one operating reactor and another nearing completion carried on to develop their own CANDU type reactor until 2008 when the United States entered an agreement to support nuclear power development.

Table 6.3 clearly shows how India has developed its nuclear industry on its own. The first reactors were two 150 MW BWR

TABLE 6.4: INDIAN NUCLEAR POWER PROGRAMME

Reactor	*State*	*Type*	*Output MWe*	*Service date*
Operating				
Tarapur 1 & 2	Maharashtra	BWR	2 x 15	1969
Kaiga	Karnataka	PHWR	4 x 202	2011
Kakrapar	Gujarat	PHWR	2 x 202	1995
Kalpakkam	Tamil Nadu	PHWR	202	1986
Narora 1 & 2	Uttar Pradesh	PHWR	2 x 202	1992
Rajasthan 1	Rajasthan	PHWR	90	1973
Rajasthan 2	Rajasthan	PHWR	187	1981
Rajasthan 3 & 6	Rajasthan	PHWR	4 x 202	2000
Tarapur 3 & 4	Maharashtra	PHWR	490	2006
Under Construction				
Kudankulam	Tamil Nadu	AES-2006	2 x 1000	2017
Kalpakkam	Tamil Nadu	FBR	500	2012
Kakrapar 3	Gujarat	PHWR	2 x 700	2016
Planned				
Rajasthan	Rajasthan	PHWR	2 x 700	2017
Kudankulam	Tamil Nadu	AES-2006	6 x 1050	2021
Jaitapur	Maharashtra	EPR	6 x 1700	2022
Kaiga	Karnataka	PWR	1000	2012
Kumharia	Haryana	PHWR	4 x 700	
Bargi	Madhya Pradesh	PHWR	2 x 700	2012
Kalpakkam	Tamil Nadu	FBR	2 x 500	2020
Rajauli	Bihar	PHWR	2 x 700	
Oswara	Rajasthan	PHWR	2 x 700	
Markandi	Orissa	PWR	6 x 1000	
Mithi Virdi	Gujarat	AP1000	6 x 1250	2020
Pulivendula	Andhra Pradesh	PHWR	2 x 700	
Kovvada	Andhra Pradesh	ESBWR	6 x 1350	2021
Nizampatnam	Andhra Pradesh		6 x 1400	
Haripur	West Bengal	AES 2006	4 x 1050	2023
Chutka	Madhya Pradesh	PHWR	2 x 700	

installed by GE at Tarapur in 1969, the next reactor was a 220 MW CANDU plant by Atomic Energy of Canada. The intention was to standardize on CANDU because it had been designed to also operate on a Thorium fuel cycle. Of currently estimated world resources of Thorium, approximately 2.6 million tons, India holds 19% in a large deposit in the south in Tamil Nadu state.

Until completion of Tarapur in 2005, with two 490 MW PHWR,

these were the only reactors that India built and were all rated at about 200 MW. But the development continued and the two units at Tarapur are being followed by 700 MW units, of the first two will be at Kakrapar in Gujarat.

Now the first foreign supplied PWR's are under construction at Kudankuam in Tamil Nadu. The two Russian-designed AES-2000 are scheduled for completion in 2016 and 2017. Ultimately there will be eight of these reactors on the site. All nuclear stations to date have been built by the Nuclear Power Corporation of India Ltd (NPCIL). But National Thermal Power Corporation, the country's largest generating company, is also planning to build a 2000 MW nuclear plant in either Haryana or Madhya Pradesh.

NPCIL however has decided to group the foreign reactors in nuclear energy parks with six or eight units giving 9-10,000 MW each. In addition to Kudankulam others are planned at Jaitapur, Mithi Viridi, Kovvada and Haripur.

Of these Jaitapur will have six of the Areva 1600 MW PWR for a total of 9600 MW. Mithi Viridi and Kovvada will both have American reactors with the eight Westinghouse AP 1000 on one and GE's ESBWR at the other. Haripur is expected to be a second Russian site with AES-2000 Reactors.

From 1990 to 2008, electricity production trebled in India to 830TWh, but the wide separation of generating plants from their loads meant that nearly a quarter of it was taken in transmission losses so that consumption was only 591 TWh equivalent to about 700 kWh per capita. Of this 68% was from coal, 14% from hydro, 8% from gas and only 2.5% from nuclear, mainly because of a shortage of natural uranium. Nuclear power since 2008 has increased with uranium imports as well as greater domestic supply. The drive to nuclear energy aims to achieve 25% of supply by 2050. The country has low coal and gas reserves and the big hydro reserves are up north in the Himalayas.

Indeed the majority of nuclear power development for the next few years will be in Asia, specifically in China and India, but also Vietnam, Thailand and Malaysia. These are all countries with low rate of electricity consumption and growing populations, and wide disparities in wealth, and are looking at initial orders for one or more 1000 MW PWR's. The largest nuclear power programs,

TABLE 6.5: CURRENT CHINESE NUCLEAR PLAN

Power plant	*Province*	*Output MW*	*Reactor type*	*Service date*
In operation				
Daya Bay	Guangdong	2 x 1000	PWR	1994
Qinshan 1	Zhejiang	279	CNP-300	1994
Qinshan 2	Zhejiang	3 x 610	CNP-600	2010
Qinshan 3	Zhejiang	2 x 665	Candu 6	2003
Ling Ao 1	Guangdong	2 x 1000	PWR	2003
Tianwan 1&2	Jiangsu	2 x 1000	VVER-1000	2007
Ling Ao 2	Guangdong	2 x 1000	PWR	2010
Under construction				
Ling Ao 3	Guangdong	2 x 1037	CPR-1000	2011
Qinshan 4	Zhejiang	650	CNP-600	2012
Hongyanhe	Liaoning	4 x 1080	CPR-1000	2014
Fangjiashan	Zhejiang	2 x 1080	CPR-1000	2014
Sanmen	Zhejiang	2 x 1250	AP1000	2014
Taishan	Guangdong	2 x 1770	EPR	2014
Tianwan	Jiangsu	2 x 1060	VVER 1000	2013
Xudabao	Liaoning	2 x 1250	AP1000	2014
Sanmen	Zhejiang	2 x 1250	AP1000	2014
Haiyang	Shandong	2 x 1250	AP1000	014
Xiaomoshan	Hunan	2 x 1250	AP1000	2014
Fuqing units	Fujian	2 x 1080	CPR1000	2014
Ningde	Fujian	6 x 1080	CPR1000	2015
Haiyang	Shandong	2 x 1250	AP1000	2015
Hongyanhe	Liaoning	2 x 1080	CPR1000	2015
Changjiang	Hainan	2 x 650	CNP600	2015
Hongshiding	Shandong	2 x 1080	CPR1000	2015
Shanwei	Guangdong	2 x 1080	CPR1000	2015
Shidaowan	Shandong	210	HTGR	2015
Xianning	Hubei	2 x 1250	AP1000	2015
Wuhu	Anhui	2 x 1250	AP1000	2016
Yangjiang	Guangdong	4 x 1080	CPR1000	2016
Fangchenggang	Guangxi	2 x 1080	CPR1000	2016
Fuqing	Fujian	6 x 1080	CPR1000	2016
Pengze	Jiangxi	2 x 1250	AP1000	2016
Taohuajiang	Hunan	4 x 1250	AP1000	2018
Sanming	Fujian	2 x 880	BN800	2019

in terms of number of units, are in China and India; and for percentage of nuclear power for electricity production are the big industrial countries of the Asian Pacific, Japan, and Korea.

Secondly there are at least three Asian countries which have developed the complete fuel cycle China, India and Japan to which could be added the Russian Federation, because this is potentially a big market for Rosatom and not just for the AES-2006 PWR. For Russia has also developed the Fast Neutron Reactor (FNR) and has sold the first two examples of the BN800 to China.

If you control the whole fuel cycle there are two things that can be done to ensure future energy security. First the plutonium and uranium recovered from the reprocessing of the spent fuel can be made into mixed oxide fuel and returned to the existing reactors. Second, with FNR the plutonium can be put in the core with natural uranium in the blanket to make more plutonium for mixed oxide fuel and to use in another FNR to consume the higher actinides from the original fuel charges.

China has also developed a small high-temperature, gas-cooled reactor (HTGR), of which the 10 MW prototype, was started up in in 2007 at Tsinghua University, north of Beijing. The pebble-bed modular reactor design was originally developed in Germany in the 1970s and a prototype was built at the Julich Nuclear Energy Research Centre, near Aachen. When development stopped under Green anti-nuclear presssure the technology was sold to a new company, PBMR Pty Ltd in South Africa, set up in 1996.

PBMR was working in partnership with Westinghouse and funded by the South African Government. The company designed a 400 MW plant and got as far as ordering components and making the impregnated carbon spheres which form the fuel. Samples were sent for testing to the Idaho National Laboratory, and to the Russian Nuclear Laboratory at Ekaterinburg.

But in 2010 after putting over ZAR 9 billion ($1300 million) into the project over eleven years thc Government announced that it would stop funding and buy PWR for public electricity supply. There are already two 900 MW units running at Koeburg, north of Capetown. Westinghouse pulled out and their work is now one design in the Next Generation Power Reactor.

China is now building a commercial prototype HTGR at Shidaowan, in Shandong province which is planned to enter service in 2015. This 210 MW plant will generate electricity, but the real

6.9 Civaux, France: this station on the River Vienne, south of Poitiers has the third and fourth of the 1450 MW type N4 reactors from which was developed the EPR now under construction in Finland and France. (Photo courtesy of EdF)

benefit of the HTGR is its high operating temperature. Given that so many industrial processes burn fossil fuels for process heat at high temperatures. A small reactor with no emissions would be attractive to these industries in the future when there is a viable price for carbon emission which they would be obliged to add to the cost of their products.

These same arguments apply in Europe where the political will is to have a carbon emission price which would encourage interest in renewables and nuclear. But first every nuclear reactor that can should get a life extension to sixty years because renewables are not going to supply the high temperature process heat for industry, not even solar, and many of the processes are essential to social wellbeing.

The cost of nuclear power has been manipulated by the antics of anti-nuclear factions arguing at public hearings and extending the time before any agreement is reached, equipment can be ordered and the first concrete poured. The industry has therefore concentrated on the simplification of their designs to lower the cost and shorten the time of construction.

The problems back in the early years were that there were few

6.10 Olkiluoto, Finland: construction site for unit 3 in June 2011. The reactor building is complete and installation of reactor and steam turbine is underway for completion at end of 2012. (Photo courtesy of Teollisuuden Voima Oy)

standard designs. The latest designs have addressed this problem and established a form of license which could be applied anywhere. In March 2011 in the United States, GE received from the Nuclear Regulatory Commission the final safety evaluation report (FSER) and final design approval (FDA) for their ESBWR (Economically Simplified Boiling Water Reactor).

The FDA proclaims that the ESBWR design is safe, that all technical issues have been resolved. This means that the reactor can be built anywhere in the world that recognises the FDA for a reactor design and its acceptance by the country of origin, i.e. the United States. However, in the home country it is a last step before final design certification, which for ESBWR was granted in the autumn of 2011.

So what of Europe and the United States? First there are links to industry on both continents. Westinghouse has already linked to three companies in the United Kingdom, and has also formed an association with the Spanish utility ENDESA in preparation for future orders in Spain and South America where Spain's heavy industries will be able to manufacture reactor vessels and steam generators for export.

In the United Kingdom the Government announced a floor price for carbon emissions in April 2011 which has given a cost advantage to nuclear power. Areva have said that their first two EPR reactors will be installed as Hinkley Point C in Somerset, where there are an existing 1320 MW B station which it is planned to close in 2016, and the A station with two Magnox reactors which is being decommissioned.

The new Hinkley Point C reactors will be rated at 1700 MW and will an important electricity generator for the south west of England. When the B station shuts down, the only power stations still running in the southwestern counties will be Centrica's 900 MW combined cycle at Langage, near Plymouth completed in 2009, and a single gas turbine peaking plant which replaces an earlier plant at Indian Queens, Cornwall.

France is currently building an EPR at Flamanville and has earmarked another site on the Channel coast at Penly. The long term plan is to replace all of the 900 MW reactors with new EPR's.

The first nuclear reactors for power generation were developed by the victorious countries of the Second World War which was brought to an end by dropping a plutonium bomb on the Japanese city of Nagasaki. Atmospheric testing of bombs continued into the 1950s by the United States Soviet Union, United Kingdom, and France until an international agreement forced all weapon testing to be conducted at underground sites. For a bomb you needed plutonium so you had to have a fission reactor producing enough fast neutrons to make it, which meant enriching the uranium fuel. Then when you had used up the fuel, you needed a reprocessing plant to separate the plutonium.

The subjugation to Communist government of the countries which the Soviet Union had liberated from Germany; the Berlin blockade which was overcome by an eleven-month airlift from June 1948 to May 1949, defined the start of the Cold War. Bomb tests showed to each side that each meant business if the other put a foot wrong.

This lasted for nearly 50 years until the Chernobyl accident forced the Soviet Government to accept help with its nuclear power plants shortly after which the Communist Governments of

Europe one by one collapsed under public pressure. Finally the Soviet Union itself broke up into its seventeen republics, three of which are now in the European Union.

Against this historical background nuclear power developments continued. Eight countries have developed nuclear reactors but not all are producing them today. Of the four producers in the United States, three have developed Pressurized Water Reactors (PWR) and one has developed the Boiling Water Reactor (BWR).

In the United Kingdom development concentrated on gas-cooled reactors, with first the natural uranium Magnox design, followed by the Advanced Gas-Cooled Reactor (AGR) the Fast Neutron Reactor, and a prototype High Temperature Gas-Cooled Reactor (HTGR).

France similarly started with gas-cooled reactors but had taken a license from Westinghouse for the PWR. They also developed Fast Neutron Reactors. They also had found uranium deposits in the Massif Central and in their desire to take oil out of power generation decided to go for the PWR in a standard 900 MW size, of which the first two were installed in Alsace at Fessenheim and went into operation in the winter of 1977-78.

Sweden has developed an innovative BWR which has had a major influence on GE's design for the ABWR. But the last BWR to be supplied from Swedish industry was the 1170 MW unit 3 at Forsmark, on the Baltic Coast 100 km north of Stockholm in 1985.

Canada's unique contribution to nuclear technology is the CANDU reactor which is a heavy water reactor in which natural uranium fuel is packed in horizontal pressure tubes passing through the moderator tank which is known as the Calandria. There are twenty CANDU reactors in Canada of which 18 are in Ontario and one each in Quebec and New Brunswick.

Russia was among the earlier developers and produced the RBMK design to produce plutonium for its weapons programme, and a PWR, which in its initial 430 MW version, was sold in the Communist countries of Eastern Europe and to Finland, where the two units at Loviisa, were modified with a Western containment and control system to bring them up to the safety standards of the nuclear plants in North America and western Europe.

6.11 Palo Verde, AZ: the largest nuclear power plant in the United States with 3 x 1300 MW PWR. The steam generators are the design base for the Westinghouse AP1000. (Photo courtesy of Arizona Public Service)

China is the last country to develop a nuclear reactor. The first in service was a 300 MW PWR designed with some help from Mitsubishi Heavy Industries who also manufactured some of the reactor hardware. China Nuclear Power Engineering (CNPE) is now concentrating production on the CPR 1000 a 1080 MW PWR of which the first was installed as Quinshan unit 5 and went into operation in 2004. Another eight of these reactors are under construction for entry into service before the end of 2015.

Other builders of reactors are among the original licensees and constructors of the American reactors in Europe and the Far East who have continued development into new designs. In France, Framatome, now incorporated in Areva developed the 900 MW design for Electricité de France and later enlarged it to 1300 MW. Both designs have since sold abroad in South Africa and Korea.

Germany similarly took American licenses with PWR going to Siemens and BWR to AEG. Both later merged their power generation activities into Kraftwerk Union and completed seventeen reactors, eleven of which are PWR and the remainder are BWR. The last was at Neckarwestheim which went into operation in April 1989. Since then German public opinion has

6.12 Tihange, Belgium: Westinghouse completed the three reactors between 1975 and 1985. Third unit in foreground has been used to test reactor internals for the new AP 1000. (Photo courtesy of Electrobel)

swung completely against nuclear energy, though Siemens is still building steam turbines and generators for nuclear plants.

Westinghouse has a nuclear policy of using local firms as much as possible to build components such as reactor vessels and steam generators and construct the plants. The company's European Headquarters are in Belgium where they have built all the seven reactors currently in operation there. They have also built Beznau in Switzerland, the country's first nuclear power station and is being considered for two new reactors being planned.

In the Far East Mitsubishi Heavy Industries took a Westinghouse license for PWR and Hitachi a BWR License from GE. Japan has an unusual electricity supply system, with Hokkaido and the north of Honshu running at 50 Hz; and the south of Honshu, and the islands of Shikoku and Kyushu running at 60 Hz, with high voltage direct current links across the boundary south of Tokyo.

Mitsubishi are mainly in the south of the country and Hitachi are in the North, most of the PWR plants are in the 60 Hz area with the majority of BWR in the north. Of the 55 reactors operating, at the end of 2010, 23 are PWR, 31 are BWR of which the Tokyo Electric Power Company (TEPCO) have 17 BWR on three sites.

TABLE 6.6: REACTOR SUPPLIERS TO THE WORLD

Company	Country	Reactor		Output MW
AECL	Canada	PHWR	ACR1000	1200
Areva	France	PWR	EPR	1600
		PWR	ATMEA	1100
		BWR	KERENA	1250
Babcock & Wilcox	USA	PWR	mPower	125
CNPE	China	PWR	CPR 1000	1000
		HTGR		210
GE Hitachi	USA	BWR	ABWR	1350
		BWR	ESBWR	1600
Doosan	Korea	PWR	OPR 1000	1000
		PWE	OPR 1400	1350
Hitachi	Japan	BWR	ABWR	1350
Mitsubishi	Japan	PWR	USPWR	1700
Rosatom	Russia	PWR	AES 2006	1080
		FNR	BN 600	600
Westinghouse	USA	PWR	AP 1000	1100
		PWR	SMR	200

In 2006 Toshiba took over Westinghouse. ABB, before it joined with Alstom, had taken over Combustion Engineering except the nuclear operations which were sold to Westinghouse. In Korea the Combustion Engine System 80 design is the basis of two new designs of PWR. The first two System 80 reactors were Yonggwang units 5 and 6 after which the design was metamorphosed into the Korean Standard Nuclear Plant for export to Asia, principally to Vietnam and Indonesia, but their first contract is from Abu Dhabi.

The first examples of the Generation-III reactors have been in commercial operation for more than ten years. These are the GE ABWR's of which the first went into operation with TEPCO as units 7 at their Kashiwazaki Kariwa site in November 1996, and was followed seven months later by the second unit. All told there are four in operation, with three under construction in Japan, and two in Taiwan.

Areva's first EPR's at Olkiluoto, Finland, and Flamanville, France will be going into service in the winter of 2012-2013. Construction of a third at Penly, near Dieppe will start in 2012. The first examples of the Korean Standard Nuclear Plant OPR 1000

TABLE 6.7: FIRM NUCLEAR PLANS IN EUROPE

Country	Unit	Output MW	Reactor type	Service date
Bulgaria	Belene 1	1060	PWR	2016
	Belene 2	1060	PWR	2017
Czech Republic	Temelin 3	1200	AP 1000	2023
	Temelin 4	1200	AP 1000	2024
Finland	Olkiluoto 3	1720	EPR	2013
	Olkiluoto 4	1720	EPR	2020
	Pyhajoki	1300	EPR or AP 1000	2021
France	Flamanville 3	1750	EPR	2016
	Penly 3	1750	EPR	2017
Lithuania	Visagenas	1350	ABWR	2020
Netherlands	Borselle 2	1600	EPR or AP 1000	2018
Romania	Cernavoda 3	750	CANDU	2016
	Cernavoda 4	750	CANDU	2017
United Kingdom	Hinkley Point C1	1700	EPR	2018
	Hinkley Point C2	1700	EPR	2019
	Sizewell C1	1700	EPR	2021
	Sizewell C2	1700	EPR`	2022

have been in operation since 1998. Russia has had examples of the VVER 1000, from which the AES 2006 is derived, in operation since 1986.

Westinghouse has followed a different design approach by testing components in customers' existing reactors. It has already built several reactors of more than 1000 MW electrical capacity and designs of the basic AP 1000 hardware have been applied to some of these earlier reactors at Palo Verde and Tihange.

The major difference in the APR 1000 is that this is a two-loop design instead of the four-loop design of the earlier models. Each steam generator is based on the design of those at Palo Verde, AZ. Similarly the reactor and its internals are based on the design of Tihange 3 in Belgium.

Palo Verde is located about 70 km west of Phoenix, AZ, and was completed in 1988. It has three 1270 MW PWR's by Combustion Engineering, and is the largest nuclear plant in the United States. An unusual feature of the plant, since it is on a desert site, is that the cooling water is the treated sewage effluent of Phoenix and other towns in Arizona. It is an important hub in the western States

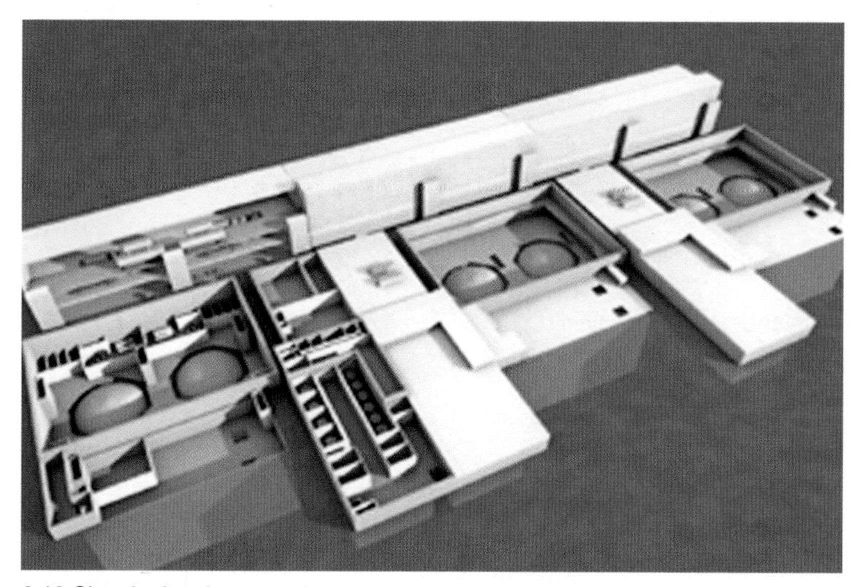

6.13 Sketch showing a station with six mPower reactors for a total capacity of 750 MW. The reactors are underground and the tall building behind contains the steam turbines and control room. (Photo courtesy of Babcock & Wilcox)

electricity supply system with 500 kV lines going out to San Diego and Los Angeles.

Of the new reactor designs available four have outputs below 1000 MW. These include the 200 MW Chinese HTGR based on the original German pebble bed reactor. Scaled up from the 10 MW prototype at Tsinghua University, construction started in March 2011 with planned entry into operation in 2015.

Similarly the Russian BN600 Fast Neutron Reactor follows the design of a prototype installed as Byeloyarsk 3 and which went into operation in November 1981. Twenty-eight years later, it was shut down for maintenance and upgrades to the steam turbine and control system and has received a 15-year life extension to 2025.Two more BN600 are in production for export to China.

Two small reactors, the Babcock & Wilcox mPower rated 125 MW and the Westinghouse SMR at 200 MW are of similar design. The letter M in each name stand for modular, which implies that growth in electricity demand could be met as in earlier times by having say four units in a station, each installed twelve months apart. The first unit would enter service in, say, November 2019 and the last in November 2022.The particular feature of both designs is

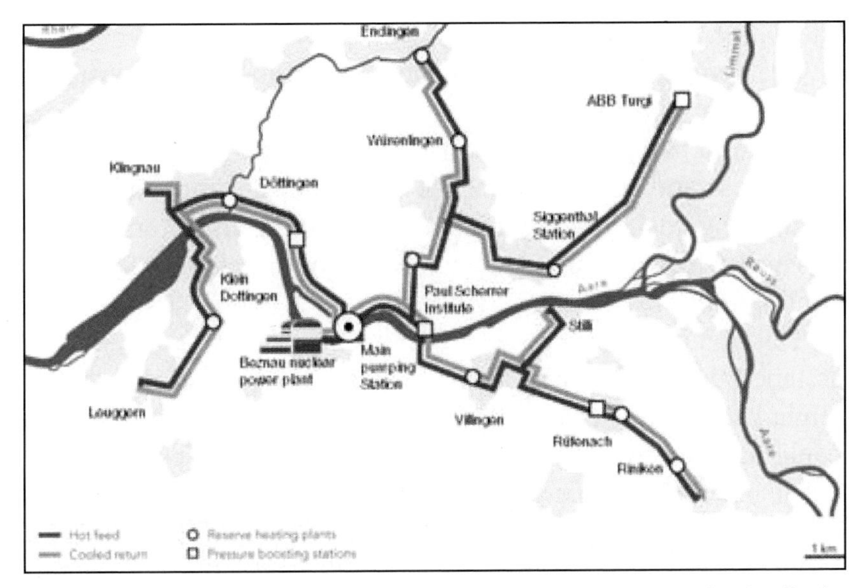

6.14 Beznau, Switzerland: nuclear plant has been the heat source for the district heating system serving eleven communities in the lower Aare Valley since 1984. (Diagram courtesy of Refuna)

that there is a single tall pressure vessel containing the pressurizer in the top of the vessel with below it a ring of heat exchanger modules mounted above the reactor. The only penetrations are for the condensate input to and steam output from the boilers and their individual circulation pumps.

But until now the typical nuclear power plant is a PWR of around 1000 to 1200 MW powering a single large steam turbine and with a reactor which must be refuelled annually. They are all installed in large industrial countries with heavily reticulated transmission systems that can operate such a large unit.

The small reactors are designed to be small enough to build in a factory and which can be shipped as a complete unit on a train or by road. Babcock & Wilcox have joined with Bechtel to form a new company, mPower Generation, who plan to offer the reactor as a turnkey contract.

The small reactor is designed to appeal to the small electricity supply systems which cannot accept a 1000 MW unit but want the clean energy supply that nuclear power can provide. Then there are the utilities with low load growth that can install large gas turbines at regular intervals the small reactors which could be a viable alternative for base load power.

The complete power station could have four reactors installed at intervals, and each driving a steam turbine, both reactors are designed to be installed underground with just the steam turbines switchgear and control room above. This is an important factor in lowering cost, because it dispenses with the large containment structure which all current reactors have and which must be able to withstand an aircraft crashing into it.

This requirement was set up more than thirty years ago and assumed that it would be a small private or military aircraft that might be in difficulty in bad weather with low visibility. Nobody could have then envisaged it being a deliberate attack by a fully laden, hijacked commercial airliner.

Global warming has given the boost to nuclear energy that it badly needed. With the average per capita electricity consumption in northern Europe 6000 kWh and more than double in North America, it is clear that conservation and renewables cannot give a reliable electricity supply on their own. If coal and oil are taken out of the power generation market, gas hydro and renewables are the complementary fuels.

Many countries in the Far East seem to have got the message with China and India leading, Korea is not only building for itself but has secured a first export order for four PWR of a design which it is building at three sites.

In Europe there are differences from east to west both in the number of reactors being built and their choice of equipment. The reactor choice seems to be between two PWR types, the EPR and the Westinghouse APR 1000. However Lithuania has chosen the ABWR of which several are already operating in Japan; and GE has also made arrangements with Energoproject Warszawa SA in Poiland to seek out possible subcontractors for the country's nuclear projects. Polska Grupa Energetyczna is studying western reactor designs with a view to buying two for service in about 2023.

All European plans are for the large reactors but Finland, alone in Scandinavia and in Western Europe, France, the Netherlands and the United Kingdom have any firm plans for construction among the nine nuclear operating countries.

Table 6.8: SMALL REACTORS UNDER DEVELOPMENT

Name	Output MW	Type	Developer	Comments
International				
IRIS	335	PWR	Westinghouse	Integrated reactor design
GT-MHR	285	HTGR	General Atomics	Gas cooled GT output
FUJI	100	MSR	ITHMSO	Sodium cooled fast reactor
Argentina				
CAREM	200	PWR	CNEA & INVAP	Construction Authorised
China				
HTR-PM	210	HTGR	INET & Huaneng	Completion scheduled 2015
Russia				
KLT-40S	35	PWR	OKBM	2 on floating nuclear plant
VK-300	300	BWR	Atomenergoproekt	Designed for CHP schemes
BREST	300	FNR	RDIPE	Lead-cooled reactor
SVBR-100	100	FNR	Rosatom/En+	Pb-Bi cooled fast reactor
South Korea				
SMART	100	PWR	KAERI	Integrated reactor design
United States				
SMR	200	PWR	Westinghouse	Integrated reactor design
mPower	125	PWR	Babcock & Wilcox	TVA order for Clinch River
Hyperion PM	25	FNR	Hyperion Power	Pb-Bi cooled fast reactor
NuScale	45	PWR	NuScale Power	Designed for CHP schemes
Prism	311	FNR	GE-Hitachi	Sodium cooled fast reactor

But the smaller reactors have a key role to play. First they are designed for factory assembly and installation underground. Some will have higher enrichment fuel which will have to be changed perhaps every five years. All of this suggests a plant which will be cheaper to build and operate.

One application for small reactors is on Island networks where electricity is still generated with oil. A group of small reactors could provide a clean and flexible energy supply. But small reactors will also be needed if nuclear energy is to be applied to a wider range of applications where the reactor must be located closer to the point of use. A small reactor would be less than the 300 MW capacity of a large gas turbine, such as would supply power and steam to factories on a large industrial estate, or provide district heating to a large town.

In fact there has been some nuclear combined heat and power already. The first nuclear plant in the UK at Calder Hall, Cumbria supplied steam to the Sellafield reprocessing plant for forty years.

Since it was shut down, Fellside Cogeneration, a 180 MW gas fired combined heat and power plant has taken over to supply steam.

Refuna is the District heating company serving the lower Aare Valley in Switzerland. The network has been developed since 1984 and is supplied with heat from the Beznau nuclear plant. The network is 130 km long and serves eleven towns in the region with some 15,000 individual customers for heat, including some research institutes and commercial premises northwest of Zurich.

Further upstream on the Aare is Gosgen Daniken a 985 MW PWR built by Siemens which went into operation in 1979. This supplies steam to a nearby paper mill. Elsewhere in Europe there is nuclear district heating in the Czech Republic and Slovakia.

The Bohunice nuclear plant in northwest Slovakia has four Russian PWR built in two phases. Units 1 and 2 were shut down in 2006 and 2008 and units 3 and 4, which are VVER 440 213 rated at 472 MW, were put into operation in 1984 and 1985. At about the same time a district heating scheme was being installed in the nearby town of Travna and the steam turbine of unit 4 was modified by installing three district heat exchangers, so that it could act as source of heat for the scheme.

The present station is due to be closed in 2025, by which time a new reactor, either a Westinghouse AP 1000, or an Areva EPR will have been installed as Unit 5 to take over the district heating for the following 60 years.

District heating from nuclear plants was first talked about in Communist times in Czechoslovakia, before the country split into the Czech Republic and Slovakia. At that time nuclear energy was only just starting and two Russian-designed VVER 400-230 PWR were being built at Bohunice in what is now northern Slovakia. These were shut down in 2008 and 2009 as a condition of Slovakia joining the European Union. But two of the later model VVER 400-213 had been installed in 1984 and 1985, and at the same time the four unit Dukhovany plant, with the same reactor type, was completed in the Czech Republic.

Nuclear distict heating, except for the Swiss Refuna scheme has been an East European interest with the Czech Republic, and Slovakia starting after the Communist era. But a somewhat older scheme is at Cernovoda, Romania. The plant is located to the south of the city, on the east bank of the Danube Black Sea canal.The original plan was for five 720 MW CANDU reactors of which two were completed in 1996 and 2007 respectively. The site is connected to the city district heating scheme and supplies some 170 GJ/year.

In Finland there are extensive district heating systems with the largest in Helsinki covering 96% of the built up area. The country was the first in Europe to start construction of a new nuclear plant in order to cut their greenhouse gas emissions. With unit three at Olkiluoto coming into service in the winter of 2012-13, the other nuclear operator, Fortum, wanted the next unit to go on their site at Loviisa, some 130 km east of the capital. Their plan was to run a district heating trunk main to the eastern suburbs of Helsinki.

This has been thwarted by the government's decision to site the next reactor, the country's sixth, at Olkiluoto. The planned seventh unit will go in the north near Pori. Given the age of the existing coal-fired plants and that the last to be built, the 565 MW Meri Pori station was completed in 1994, it would come as no surprise if Finland were in fact aiming to take coal out of power generation.

District heating with nuclear plant is suitable because of their relatively low steam temperatures. Another similar application is the distillation of sea water for public water supply. Could this be one of the reasons for developing small reactors to provide a system which could be installed closer to the site of application.

Of the fifteen reactors listed in Table 5.6 above, seven are PWR ranging in output from 300 down to 25 MW, four are fast neutron reactors and two are high temperature gas-cooled reactors. Of these the only one under construction is the Chinese Pebble Bed Modular reactor which is planned for operation in 2015; and the Russian KLT 40 S PWR of which two have been supplied to the first of a series of floating power barges.

General Atomics 300MW gas-cooled reactor is their first since their unsuccessful attempt to launch large high-temperature gas-cooled reactors during the 1970s. At the time the Nuclear Regulatory Commission was struggling against fierce anti-nuclear protests in licensing the water-cooled reactors. Despite the intrinsic safety of the gas-cooled reactor and its higher efficiency there was no chance of getting it licensed at that time.

But the fast reactors which are the most interesting because of the nuclear clean-up role envisaged for them. Already Russia has sold two of the larger sodium-cooled BN800 to China for that purpose. For clean up the reactors will have a plutonium core and a spent fuel blanket. Fast neutrons from the plutonium impact the blanket and the higher actinides, which are produced in small quantities in a PWR reaction, are consumed in the blanket reaction to produce more electricity.

Not all fast reactors are sodium-cooled. The Russian BREST reactor has a liquid lead coolant, while the SVBR 100 and the 25 MW FNR of Hyperion Power, have a lead-bismuth coolant. Neither of these coolants exhibit a violent reaction with water and solidify at a lower temperature than sodium.

Hyperion Power of Los Alamos, NM, have the smallest reactor which is factory sealed with a uranium nitride fuel. The only external connections are to a steam generator which outputs at 500°C. The refuelling interval is about 10 years and the method will be to send the reactor module back to Hyperion. Since it measures only 1.5 m diameter by 2.5 m high it can be sent by road or rail from an American customer and flown from anywhere else in the world. By reducing the temperature so that the coolant solidifies it can be transported safley around the world.

Small reactors have yet to be licensed for application, at least in the United States, and the first are likely to be the PWR's. In fact Tennessee Valley Authority has issued a letter of intent for two mPower 125 MW units to be installed at their Clinch River site by 2020.

All but two of the small reactors are designed to be installed underground which removes the need for the massive containment structures of the present reactor designs. The big question is whether the public would accept an underground nuclear system closer to their homes or places of work. Or because it was out of sight would it be out of mind? For this is the only way we can apply nuclear energy to a wider range of applications to industry and public service.

7
Why not nuclear merchant ships?

Shipping has been an essential transport industry for all of recorded history. For all but 200 years ships on the high seas were propelled by sails, susceptible to powerful storms which could sink or wreck them, or equally to lack of wind which could becalm them for days on end.

Ships powered by steam or diesel engines can ride out the worst storms and cannot be becalmed. Today they can carry people and goods all over the world though personal international travel is mainly by air, and sea travel is on ferry services or on cruise liners for a holiday.

From about 1970 with the arrival of the wide-bodied turbo-fan powered aircraft, international leisure travel has increased rapidly. The great ocean liners of which one of the most famous was the *Queen Mary*, and which is now permanently moored as a museum ship in Long Beach, CA, could not compete on time or cost and were retired or converted into cruise ships for a broader class of passengers. But a steady increase in the price of oil and higher fares boosted by passenger taxes and fuel surcharges have in recent years reduced airline passenger numbers.

However, the freight carriers have had to pay the higher fuel prices and their ships are now too big to return to sail power. So nuclear is the clear answer. Nuclear surface ships have been constructed besides including the Russian ice-breakers and the container ship *Sevmorput* which has ice-breaking capability supplies the arctic communities.

Freight traffic is the basis of a country's trade in manufactured goods, raw materials, and fuel which will continue long into the future. How these ships are propelled is therefore a challenge for

7.1 NS Savannah was the first nuclear powered cargo ship to be built but at the wrong time. Operating costs were too high and the much larger super tankers and container ships were starting in service.

the future.

The United States, Russia, and the UK were the first to apply nuclear rectors to, initially submarines, which is logical because they can stay below the surface for weeks on end and do not have to come up to the surface periodically to recharge batteries. Besides submarines, aircraft carriers and military surface ships have been built and many remain in service, but although there is a considerable experience of operation and maintenance of reactors on the high seas, only three cargo ships have been built, and none remain in service. But for more than twenty years every winter a fleet of Russian nuclear ice-breakers have worked along the arctic coast.

Sixty years ago in the United States Dwight D. Eisenhower was elected President. Only seven years had passed since the ending of the Second World War with the nuclear bombing of Hiroshima and Nagasaki, and with the continuation of atmospheric testing, there were few people who thought that the adjective nuclear could be applied to anything other than a weapon.

The decade of the fifties would see the first nuclear power plants built in the UK, France, the Soviet Union and the United

7.2 Russian ice-breaker *Rossia* at work in the Arctic winter. Two of seven such ships which have been escorting large tankers through the ice in the summer months as a short cut to China and Japan. (Photo courtesy of Atomflot)

States; and also the first nuclear submarines. But Eisenhower wanted to expand the peaceful uses of nuclear energy and saw a nuclear powered cargo ship which could travel around the world as an ideal demonstration.

The *NS Savannah* was launched from the North American Shipbuilding Corporation's yard in Camden, NJ, in 1959. It was designed as a passenger and cargo ship with a streamlined hull and was completed in 1961. It was 181.66 m long by 23.77 m beam with a displacement of 9900 dwt. The power plant was a 74 MW pressurized water reactor by Babcock & Wilcox which powered two De Laval turbines which drove generators and a 20,300 hp propeller. It had accommodation for 100 passengers and the capacity of the cargo hold was 18,000 m^3 which could hold about 8500 t of freight.

After several years of testing *Savannah* was leased from 1965 to 1971 to American Export-Ibrantsen lines for revenue cargo service, during which time it earned the operators $12 million of revenue. But it was not profitable at a time when marine bunker fuel cost about $20/ton. There were also many changes happening in the shipbuilding industry. Japan and Korea with rebuilt shipyards and lower production costs became major players in the world.

Their super tankers and later, container ships came to dominate traffic on the high seas, and *Savannah*, for all its technical novelty cost about $2 million/year more to operate than a similar diesel-engined merchant ship at the time, and was laid up in 1972.

The Soviet Union had a greater need for nuclear ships, and in 1969 launched the ice-breaker *Lenin* at the Admiralty Shipyard in St. Petersburg. This was the first of a family of ice-breakers of which there are now seven which operate along the arctic coast of Russia including two smaller units which were built to operate in the estuaries of the big Siberian rivers. Being nuclear-powered they could stay up there all winter and did not have to go back to Murmansk to refuel.

The five large ice-breakers were built each with two 30 MW reactors for propulsion and electricity supply. Two smaller vessels each with a single reactor were built at the Wartsila yards in Finland, and delivered in 1989 and 1990, without the reactors, which were fitted in Russia at the Murmansk shipyard. The two oldest ice-breakers, *Lenin* and *Sibir* have now been decommissioned and a replacement vessel went into service in 2007.

The Arctic Fleet is an important commercial link along the North coast. The ice-breakers have a cruising capacity of 6 to 8 months and the reactors are refuelled every five years. The present fleet has seven ice-breakers and five supporting vessels, including *Sevmorput,* which is said to be the world's first nuclear-powered container ship. The fleet is owned by a division of Rosatom and operated by Atomflot out of their base in Murmansk.

In July 2011 two ice-breakers escorted a cargo of gas condensates to China. The Russian tanker *Perseverance* with a displacement of 73,000 t left the port of Murmansk June 29 carrying 60,000 t of gas condensate. It was accompanied by the ice-breakers *Yamal* and *Taimyr*. The three ships travelled at an average speed of 11 knots through the ice and after ten days were north of the New Siberian Islands and near the eastern edge of the ice in the Chukchi Sea. There the ice-breakers parted company and the *Perseverance* continued its journey alone through the Bering Strait into the Pacific Ocean.

This is a much shorter route than sending it through the Suez

Canal where they could risk being captured for ransom by Somali pirates. The Arctic route however is a case of nuclear vessels escorting a diesel powered ship which will not burn so much fuel as it would along the traditional route. The escort vessels must have high availability which diesel power does not afford them in the job that they do.

Why not nuclear-powered container ships, and other large freighter vessels? There is a tremendous experience of nuclear reactors for ship propulsion but largely based on military experience. But just as former military gas turbine operators entered civilian life as more gas turbine combined cycles and industrial combined heat and power schemes came into operation, there may surely be former nuclear submariners who can service small reactors for commercial shipping.

It would not be new technology. There are many of the small pressurized water reactors of less than 100 MW operating in the military vessels of the United States, Russia, France and the United Kingdom and it would be versions of these which would be used in nuclear container ships, large tankers, and ore carriers.

If it had been launched ten years later *Savannah* might have had a longer life at sea and might even still be operating. When it was retired in 1972 it was just one year before another Arab-Israeli war erupted and precipitated a four-fold increase in the price of oil by the OPEC nations.

In 1972 Bunker fuel oil at $20/ton gave the advantage to the diesel powered ships given the higher crew costs of the nuclear ship, and the fact that designing it as a passenger and cargo ship limited its cargo capacity. Two years later Bunker fuel oil had risen to $80/ton and the additional fuel cost was more than the previous cost advantage over *Savannah.*

By 1974, too, the shipbuilding industry had also reacted to the increased fuel cost and designed larger steam and diesel-powered ships which could carry greater quantities of cargo. But there were already large nuclear ships on the sea in the shape of the US Navy's aircraft carriers.

The largest of these ships are the Nimitz Class of ten vessels, of which the last was the *Roosevelt* commissioned in 1984. These are the largest nuclear-powered ships afloat with two reactors feeding

7.3 *USS Theodore Roosevelt,* one of the ten Nimitz class aircraft carriers. Launched in 1984 she saw her first action in Operation Desert Storm in 1991. (Photo courtesy of US Navy)

a common steam range supplying four steam turbines driving the propellers. Total shaft horsepower is 260shp, which points to a reactor capacity of about 100 MW. Overall length is 338 m with beam 252 m and a displacement of 106,000 dwt.

Of course an aircraft carrier which is effectively a transportable military base would have a much larger electrical load than the equivalent sized commercial container ship. But the reactor technology has been available for more than fifty years, and several companies now have the technology to make a small PWR.

But why hasn't the nuclear cargo ship appeared before now? Economics played a part, but so did public opinion. In 1960 design of a nuclear cargo ship was started in Germany, which was launched in 1964 as the *Otto Hahn*, named after the German Nobel Laureate credited with the discovery of nuclear fission in 1938. It was slightly smaller than *Savannah* with overall length 173 m and beam 23.4 m, but the cargo capacity was much greater at 14,040 tons. The reactor was rated at 38 MWe and produced steam at 85 bars, 300°C.

The *Otto Hahn* was commissioned in 1968, and sea trials began the same year. The first commercial voyage was to Safi, Morocco

7.4 Artist's impression of the first of eight planned 70 MW floating nuclear power plants to be permanently moored in the Russian Arctic to support economic development. (Photo courtesy of Rosatom)

for a cargo of phosphate ore in 1970. Two years later the reactor was refuelled and found to have consumed 22 kg of uranium in travelling 463,000 km. The ship continued in service for another seven years when she was taken out of service having sailed 1.2 million km and visited 33 ports in 22 countries. But it was decommissioning that presented problems.

During the late seventies, the Green movement had filtered across to Europe and had gained particular strength in Germany where pitched battles had been fought against the construction sites of some of the last nuclear power stations, and public opinion had killed the proposal for a 1200 MW plant at Whyl facing across the Rhine to France. Germany at this time was divided, with Berlin similarly divided, and 160 km inside East Germany.

Many Germans considered themselves to be on the front line of a future war, which might be waged with nuclear weapons. So to bring a nuclear ship back to Kiel or Hamburg in order to remove the reactor was totally unacceptable.

After a long search for a suitable port the Japanese Government finally offered Nagasaki. At the Mitsubishi shipyard there the reactor was removed and diesel engines put in its place. As a diesel cargo ship she continued to trade under several owners until she was sent to India in 2008 where she was scrapped at the port of Alang.

One other contribution to nuclear shipping is the floating nuclear power plant. Russia plans to build eight of these vessels which would be permanently moored at points along their arctic coast. The first of these, named Akademik Lomosovich is nearing completion in St Petersburg. In 2012 it is to be towed to the far east and moored at Vilyuchinsk on the Kamchatka peninsula where it will provide energy for the Pacific naval base, and also public water supply by distillation of seawater.

The idea of floating plants came with the break up of the Soviet Union and the sudden decline in nuclear energy business during the 1990s. It was considered that to have power plants up there to support the economy of the region would provide work for the nuclear industry. The basic power plant is 70 MW provided by two the KLT 40 S; a PWR which has been the principal power plant for their submarines and ice-breakers.

The advantage is that the plant would be fully assembled and tested in the south, before going north, and avoid the problem of having to send men equipment and machinery to build a plant in situ. Furthermore, the plant is at the centre of its own network and does not have to be connected one to the other in such a potentially inhospitable environment.

But is there another more obvious application for a floating plant on a large island which is out of range for an HVDC cable connexion from the continental mainland. Places such as Cyprus, Jamaica, and the Falkland Islands, particularly if the oil industry develops down there. Why should these people be condemned to burn oil for their heat when they have no natural energy resources that could be exploited?

General public opinion is in favour of nuclear energy as a neccessary source of electricity with no emissions; but what about nuclear powered commercial ships? The Russian ice-breakers and one of their supporting vessels, are a special case, having been designed for an extended operation in the Arctic environment. But a cruise ship or a container vessel is owned by a shipping line and registered in a particular country. Would they want to travel in one as a cruise ship? For as many people who would be curious to travel for the experience there could be many who would be scared of something happening to the reactor in a collision with another vessel or if it grounded in shallow water.

How would a future nuclear ship be crewed? *Otto Hahn* and *Savannah* each carried a large complement of nuclear engineers to service the reactor. But the rising price of oil in recent years has reduced the disadvantage of the nuclear ship. These early ships used the same reactors as the nuclear submarines, and larger reactors have been applied to aircraft carriers with their much larger electrical load on board. These reactors have been operating for more than thirty years. The only reactor lost at sea was in the Russian submarine Kursk which sank in the Baltic in August 2000. The reactor was automatically shut off, and a year later a Dutch salvage team recovered the vessel and the bodies of the 118 crew which were susequently buried in Russia.

A nuclear container ship would have a smaller crew and would have a much smaller electrical load on board. Even so the reactors could still be monitored at the vessel's home base. A number of small reactors are being developed of 100 MW or less and could be used on ships. Their particular properties of being a factory built fully integrated system with a refuelling interval of at least five years means that they could be easily installed.

The infrastructure to support a nuclear cargo ship must be at a port which can handle the vessel and refuel it. With the small reactors the enrichment is higher than for a large nuclear power station, so that the refuelling interlude can be extended to five years, as for the Russian ice-breakers. At the end of its life it could not be broken up anywhere but in a port with the equipment to remove the reactor and return it to the supplier.

The other issue which nuclear ships would challenge is the Flag of Convenience where there are a number of countries which make a business of registering ships outside of their country of ownership. Since this practice is designed to save money in registration fees and crew costs, would a nuclear ship be allowed to sail under a flag of convenience given the nuclear skills required of the engine room crew?

Then again where would refuelling take place? Would different countries' ships have different reactors so they would always have to go back to their home port for refuelling? Either that or a standard fuel design for all marine reactors must be produced so that, for example, a Korean ship could refuel in the United States, and a Brazilian ship in France.

Then there is the question of what to do with the spent fuel. Does the country performing the refuelling have the complete fuel cycle capability or does it have to send the fuel elsewhere? All these issues have to be addressed to ensure public confidence in before we can apply this new application of nuclear energy.

8 Electricity for transport

Electric tramways have been part of public transport for more than a hundred years. In fact, they predate the widespread use of cars for personal transport. In Europe electric trains have been in service as city transport, and on trunk railway lines for a similar length of time.

It was not until after the Second World War that car ownership took off and posed a real threat to long distance rail travel. The railways sought to improve their performance with clean, high speed trains, by getting rid of steam locomotives and their replacement, with diesel-electric units, and later full electrification of main lines. Diesel-electric locomotives are still widely used across North America where there are very few electrified main lines and long distance transport is mainly by air, with the railways handling mostly freight traffic.

Much of the new construction over the last sixty years has been on urban networks with extension of existing subway systems and in some cases completely new systems. The exception to this has been the Shinkansen high-speed standard-gauge rail system in Japan which started operation between Tokyo and Osaka in 1964, and the TGV system in Western Europe. The first TGV line was between Paris and Lyon, on a dedicated high-speed track which entered service in September 1981. It was first extended to Marseilles and then into Belgium, Netherlands, Germany through their own high-speed systems, and then to the UK through the Channel Tunnel.

Electrification began with DC power supplies which are still used for many urban tramways and notably throughout the London Underground network. Main line railways were initially powered

8.1 Shinkansen bound for Tokyo passes Mount Fuji. The line from Tokyo to Osaka was the first and has now been extended past Hiroshima and on to the island of Kyushu. (Photo courtesy of Japan National Railway)

by 1500 or 3000V direct current until in 1906, Norway, Sweden Germany, Switzerland, and Austria agreed on an AC system which was standardised on 15 kV, 16.7 Hz, or one third the frequency of the European electricity supply system. This network ran for most of the 20th century but has now been largely converted to the 25 kV, 50 Hz system, pioneered in France in the 1960s.

What made possible the 25 kV, 50 Hz electrification scheme was the development of semiconductors at the end of the 1940s. A few years later the high power thyristors which followed provided a compact, solid-state rectifier system.

About the same time, high voltage direct current transmission was being developed for the power companies. The initial schemes used mercury-arc rectifiers which were quickly overtaken by large multi-stage thyristor arrays. Typical applications have been to connect islands to the mainland or hydro plants built several hundred kilometres from the load centre.

All railway electrification now uses the solid state rectifiers which are operating at 25 kV which is a much lower voltage to ensure adequate clearance between the train and line side structures such as bridges, tunnels and station buildings. In fact there are

8.2 Rennes, France: TGV line is being extended from Le Mans. Until then the TGV runs on ordinary track at reduced speed for the last part of the journey. (Photo courtesy of SNCF)

twenty countries which have their entire network electrified to this standard.

The frequency means that the power can come from the mains and is another load on the power system. However it is single phase which means that individual phases insulated from each other must be connected in series. Static VAR compensators even out variations of the load on each phase as the train passes.

The Russian Federation with a wider 1524 mm track was one of the pioneers of the 25 kV 50 Hz system and the Former Baltic Republics and Finland are connected to this system. Conversion of railways between the different electrification systems continues but it depends on the importance of the route and the economic benefits that it can produce.

The most recent line completed in the UK was from Ashford, Kent to London St Pancras station, in 1996, which became the new terminal for the Eurostar Channel Tunnel services to Brussels and Paris. Previously it had terminated at Waterloo. But by moving to St Pancras they connected to all the main lines from the industrial north and also in the same street were the Euston and Kings Cross termini for lines from Scotland and northern cities.

The long debated Crossrail scheme is at last under construction and will, when completed, run from Maidenhead in the west to Shenfield via Paddington and Liverpool Street. To the west of London it will pick up the existing Heathrow electric service and to the east it will branch from Whitechapel to Shenfied, with the main route, northeast to Stratford to connect with the Eurostar services, and southeast to Abbey Wood, which passes through Canary Wharf, the business centre of the redeveloped docklands, and connects with the Docklands light railway to London City Airport.

The value of this project has been in its connexions to the other networks. Liverpool Street is the terminal for trains to Stansted Airport, While at Faringdon it will connect with the Thames link line serving the Gatwick and Luton Airports and will therefore be providing a link between Heathrow and the four other London airports.

The central section is planned to be completed in 2018 and the outer sections will follow. The Heathrow service into Paddington in the western section has been in operation since 1995. The electrification scheme is overhead 25 kV 50 Hz as on the existing Heathrow service. Maidenhead is on the main line out of Paddigton to Bristol and South Wales. This is the other project which the British government has authorised for electrification to Cardiff.

The European high-speed lines were built about thirty years ago, except for the Channel Tunnel line which entered service in 1994 to London Waterloo. In 1997 the service was moved to the redeveloped St Pancras station from which all Eurostar trains now operate. Eurostar has had some effect on air services to Paris and Brussels, and probably carries more leisure traffic than the TGV services on the mainland of Europe.

But there are no other high-speed train services in the UK as yet. However there is a plan known as HS2 which is vigorously opposed by the communities through which the likely route is planned to pass. This was chosen by the previous Government but construction is unlikely to start before 2016 at the earliest.

All high speed trains across Europe are on the Standard gauge of 1435 mm except in Russia and Finland where it is 1524 mm. But the electrification is all at 25kV, 50 Hz; similarly the Shinkansen network in Japan. The TGV system uses separate power cars with an elevated drivers cab. The Eurostar trains are similar but without the elevated cab to accommodate the British loading gauge. Shinkansen, the Chinese CRH system and most of the other high-speed systems are multiple-unit stock.

So what of future high speed trains? Attempts to build them in the United States in Florida, Texas, and California have met with opposition from airlines and other commercial interests. Of course in the United States the main purpose of the railways is to transport freight. So should the railway be a high speed or a high capacity service, shipping containers rather than people.

Russia's Trans Siberian Railway is already shipping containers from Japan and Korea into eastern Europe. From Vladivostok to Moscow is 9189 km. A freight train with containers and running at 80 km/h would reach Moscow in about 5 days. While the ship can carry more containers the train can deliver specific containers on almost a door to door service. Furthermore at the present time ships to Europe from the Far East passing through the Suez Canal run the risk of interception by Somali pirates and the crews held for ransom.

High-speed passenger trains are the transport system of a past era. Living standards and working patterns have changed and more people are living in dispersed communities far from the city centre. For many years a substantial part of daily commuter traffic into London has come from seaside towns on the south coast but mobile phones and lap-top computers allow people to work on the train in ways that they could not even twenty years ago.

High speed trains may be one solution, but to increase the capacity of the existing railway with double deck carriages and longer trains may be a more practical solution for those who use it. But for many people, given where they live and work and whether there is a railway near at hand offering the service that they would want, ownership of a car is essential. So the question is how much passenger and freight traffic can we return to the railways? Will it be because the price of gasoline becomes too high or because of the present inadequate performance of electric cars.

While there is a certain logic about using nuclear power to reduce oil consumption by ships there is only one way to reduce

8.3 Los Angeles, CA: hybrid car with a fuel cell powering the electric motors is filled with hydrogen produced at the local sewage works. (Photo courtesy of Orange County Sanitation District)

oil consumption on the roads. Indeed one hundred years ago at the dawn of motoring electric cars were quite numerous and it was not until after the First World War that gasoline-engined vehicles came to dominate the scene, since when there has been a steady development of higher performance engines and with better fuel consumption.

It is argued that the majority of people, even in North America, drive less than 150 km per day, and that therefore the ability to recharge the battery in a few hours overnight is all that is required. But it is when they have to go further that the problems start. For example your daughter, her husband and children (your grandchildren) live about 200 km away. It is an easy round trip in a day to go and see them in your present car, and you leave with a full tank and still have half a tank full left when you return home.

Then again many families go by car on holiday each year it is much cheaper than flying, but the initial journey could be several hundred kilometres. With an electric car, where are the public charging points en route, how long will it take to recharge, and if you have to do it every 160 km, how much longer will the journey take? If we can develop an electric car equivalent to a 1500 cc

8.4 Amersham, UK: engine compartment of Toyota Prius with 1500 cc engine and electronic module at bottom right which determines the use of electric motors and gasoline engine according to speed.

gasoline-engined vehicle today, and moreover, which can run at least 500 km on a full charge, then there will be a market for it. But is this the right way to go?

In the United States the emphasis is on the hybrid car. The basis of this type of vehicle is in the transmission system. The wheels are driven by individual electric motors from a battery which is charged by either an internal combustion engine or a hydrogen fuel cell. With the internal combustion engine the power supply and the drive motors are in a series parallel configuration and are able to run separately or together. In separate mode the electric motors drive and the engine charges the battery on the move. At other times the gasoline engine is supporting the electric drive motors as for instance up a hill.

Figure 8.3 shows that a fuel-cell car is filled from a special pump in much the same way as any other vehicle. The particular car in the picture has a fuel consumption of 3.4 litres/100 km or about half that of the equivalent gasoline-engined car. Hybrid cars need the battery to start the fuel cell and to power the lights and the entertainment system so the battery is charged by the fuel cell and can be used to power the drive motors. It is a similar arrangement to a normal gaoline engined car. Prototype cars have a range of about 400km.

Orange County Sanitation Distict in southern California has a treatment plant which produces hydrogen from human waste. The installation takes biogas output from the sewage plant and purifies it to hydrogen fuel. It produces 120 kg/day of hydrogen to fuel up to 30 cars and also supplies its own fuel cell which produces some 250 kWh/day for the sewage plant.

The alternative hybrid car has electric drive motors and a gasoline engine. One of the first cars of this type to be offered was the Toyota Prius which was first introduced in Japan in 1997. This had a 1.8 litre gasoline engine and two electric drive motors on the front wheels. The three units are arranged with the battery in a series parallel configuration. There was a nickel/metal-oxide battery for the electric motors which can be charged on the move by the gasoline engine. But the range on electric motors alone was only 22 km.

It is no surprise that the first hybrid cars appeared in Japan, and Korea, countries which have to import all of their oil and natural gas. The hybrids have been shown to have similar performance to gasoline-engined cars of the same size but much lower fuel consumption, but the big market for Prius has been in the western United States where more than two million have been sold.

Average fuel consumption of Prius is 3.92 litres/100 km which compares with an equivalent gasoline-engined car which would burn nearly 3 litres more to cover the same distance. The idea is that electric motors would be used in town, but outside of town at higher speeds the gasoline engine and drive motors are running in parallel with the combustion engine also charging the battery on the move. Basically the Hybrid scheme is a gasoline engined car with an electric supplement to boost output and reduce fuel consumption. Prius, in its present configuration, is a mid-sized 5-door hatchback, with a smaller 1.5 litre engine and the more powerful lithium-ion battery, which is mounted under the rear passenger seat.

Look at the websites of the motor industry and look for hybrid cars. The results will show a series of models which would look more at home on North American than on European roads, even from long established European manufacturers. Even the prices are quoted in US Dollars, rather than Euros, which suggests that they are aiming for the US market where there is a much larger vehicle population.

For every million hybrid cars, fuel consumption is effectively halved. But the fuel distribution system is the same as it has always been. For fuel cell cars the fuel is hydrogen which can be made from sewage or other waste process gas or electrolysed from water. The output from the fuel cells is water, so the fuel-cell car only uses oil for lubrication.

Electric cars in their various configurations represent a cultural change in motoring outside of North America. If you buy a new car in the United States or Canada or need to rent one, it will come with an automatic gearbox. If you want a stick-shift gearbox on your new car, you pay extra for it and may have difficulty when you eventually come to sell it.

In the rest of the world you pay extra for an automatic gearbox on a new car, and although rental companies offer automatics they charge extra for them. Electric cars are effectively automatic and people who have never driven automatics before will just have to lump it and like it.

Driving a hybrid car with a series parallel configuration of the motors, is no different from driving an ordinary automatic vehicle except for the noise. Starting out with a hill climb the main driver is the combustion engine supported by the electric motors. At the top the combustion engine reduces output and the electric motors take over. The road is flat for a few km and may then rise a few metres before it starts to descend.

The absolute speed limit in the UK is 112 km/h which de facto is in practice nearer 120 km/h. Prius is no sluggard and can keep up with other traffic. Sometimes it is running entirely on its electric motors and the gasoline engine is just charging the battery. At other times, as for instance when climbing a hill, the gasoline engine supports the electric motors.

There is a large truck in front going slower than you and as you descend the hill you occasionally brake to keep your distance. When you apply the brakes, regenerative braking returns the lost kinetic energy to the battery and your motors are gravity assisted so

8.5 The Volvo V60 is a powerful hybrid car with 2.4 litre diesel engine and lithium ion battery which gives it a hybrid range of 990 km and fuel consumption of 1.92 litres/100km. (Photo courtesy of Volvo cars)

that they draw less energy and the charge builds up. At the bottom of the hill the road is flat for several kilometres so that the electric motors are now driving the car, and the combustion engine is either supporting the electric motors or else charging the battery.

The Prius is already enjoying a strong following because it is essentially an electric car which charges its batteries on the move. In the UK the Prius costs about the same as a 2 litre gasoline engined car which has about the same dimensions and carrying capacity. The difference is that the Prius has a 45 litre tank on which it can run 965 km in combined direct drive and battery charging.

Toyota now have three hybrid models of which the 1.8 litre Auris is manufactured at their British factory in Derby. The design is similar to the Prius, with electric motors on the front wheels and a lithium ion battery.

Hybrid cars are now in production in Europe and one of the most powerful is the Volvo V60. Shown above towing a boat on a trailer up a hill which is the most convincing demonstration that hybrid cars can do everything that a normal gasoline- or diesel-engined car can do. The range is greater and the fuel consumption

8.6 Nissan Leaf is the first all electric car to go into production. But the range of 160 km can be boosted with a photovoltaic roof spoiler and regenerative braking. (Photo courtesy of Nissan)

is about one third that of the equivalent model. On electricity alone the range is only about 55 km.

There is one aspect of performance in which the electric drive motors differ from gasoline engines: regenerative braking. When an electric car brakes the motors effectively turn into generators and feed the lost kinetic energy back into the battery.

KERS, the Kinetic Energy Recovery Sytem, was introduced on Formula 1 racing cars in the 2011 season. There are two versions, electric or flywheel, which are coupled through the differential driving the rear wheels. But the system is almost instantaneous. Two cars approach a hairpin corner on the track and brake hard. The kinetic energy is stored as they enter the corner braking from 300 km/h to say, 80 km/h, and as they emerge one driver switches on the KERS before the other and receives a massive power boost allowing him to overtake the slower car.

Formula 1 is often said to be the proving ground for the motor industry. This was certainly the case with disc brakes which are now standard equipment on almost all production cars. Formula 1 KERS experience may eventually be applied to improve the fuel consumption and performance of production hybrid cars.

One of the first all electric cars in production is the Nissan Leaf, (Leading, Environmentally-friendly, Affordable, Family Car) with a range of about 170 km. Outwardly it is similar to competing gasoline-engined cars today. It is a 5-door, 5-seater hatchback with an 80 kW synchronous motor driving the front wheels and supplied from a 24kWh lithium-ion battery. Dimensions are 4145 mm long by 1770 mm wide and 1570 mm high. Distinguishing features are the absence of a conventional radiator grille and exhaust pipe. Side and tail lights are LED clusters. Wheels are 41 cm alloys. Other interior fittings and accessories are much the same as those on the equivalent gasoline-engined models.

Will an electric car be much cheaper to run? Such cars in the UK and also hybrid vehicles, do not pay the annual Vehicle Excise Duty, which for all new vehicles depends on the carbon dioxide emissions from the exhaust. Older vehicles which came before the new legislation pay a flat rate of £150. In addition, the government offer a 25% grant on the price of an electric car which is capped at £5000, and reduces the price of the Leaf to £23,600.

Nissan suggest that it will cost £260/year for a British owner to purchase electricity. On the night time tariff, currently 4.9p/kwh that is 6190 kWh providing charging from empty to full 258 times. Since the Leaf can run 170 km on a full battery the driver under these conditions could be running up to 43,860 km/year at a fuel cost of 0.24p/km. In practice few people drive more than 30,000 km in a year and many more drive only half that distance.

The equivalent gasoline-engined car with a fuel consumption of 6.5 litres/100 km at a current price of £1.35/litre the fuel cost would be 8.8 p/km, but is 650 km, on a full tank of 50 litres really comparable with the Leaf's 170 km on a fully charged battery.

In fact the Leaf may be able to travel further. An option is to fit a roof spoiler carrying a number of photovoltaic cells and which will be charging the battery during daylight hours and every time the brakes are applied even from 80 to 50 km/h small quantities of recovered kinetic energy are similarly fed back to the battery. So it is possible that the range could in practice be more than 200 km.

Initially most electric car owners would charge at home perhaps up to five nights a week depending on use, whereas the gasoline-engined car would probably fill the tank once a week. But to use electric cars in the same way as we use our present cars may prove difficult. It all depends on the extent of facilities for rapid charging and how we pay for the energy so obtained. There must either be more efficient and higher capacity batteries developed, or some means of rapid charging en route.

Electric cars are a different load which will happen on 365 days a year. Not all vehicles will need charging every day but there will be enough to present a sizeable load throughout the 24 hours, and variable from day to day. The major part of the load may be overnight but some will also require charging during the day time.

Many of the first users may buy them as second cars for local use, less than 170 km/week. Journies could be to the supermarket, a restaurant, the station, the church, or the school. Almost certainly the battery would be charged at home overnight. But eventually, with a big enough customer base there will be rapid charging points at all commonly visited places.

Experiments with wireless charging suggest that this could be applied to every parking space so that cars could be topped up whenever they entered a station, supermarket, or similar car park. To be able to do this, the banks must issue their customers with some type of parking debit card which can be used anywhere.

For example, husband and wife, go in their electric car to the supermarket, where they park it in a numbered space. A card is deposited in a reader at the exit, the pin number is entered and also the number of the parking space. The car has been parked above an induction coil which is read by the corresponding wireless reader on the vehicle underside. When the card is read the charging process starts and if the battery is fully charged, before the driver returns the process automatically stops.

An hour later the shopping is done and the same card as was used to start the charging process must be used again. The card is read to determine cost of the kWh delivered to the car and also the parking fee which are then taken from the driver's bank account. The car park is effectively the customer of the electricity supplier.

But a large conversion to electric vehicles would require more power stations to be built and a dense network of rapid charging points installed around the country. A single car can be charged off the domestic mains in 10 hours. A special vehicle charging box can do it in 8 hours. Assuming that most charging loads will be at night then this would be a 24kwh load spread over 7 hours and if a million cars were being fully charged every night that would be the equivalent output of a 3000 MW power station per million cars.

Hybrid cars have two propulsion systems, the electric drive motors and the regular gasoline or diesel engine, together with the lithium ion battery which is larger than the traditional lead acid unit since it is also driving the vehicle, typical extra weight is about 300 kg, but the faster acceleration of the electric motors, typically 0 to 100 km/h in 5.5s more than compensates. Since hybrid cars charge the batteries on the move they have a similar or longer range to that of the equivalent gasoline-engined cars.

But an electric car market will not be very large until after 2015 by which time there may be a large population of hybrid cars and the hydrogen fillers to serve them.

Electric cars, however, cannot be developed faster than the power stations needed to provide a large daily charging load. The first of the new nuclear power plants of which Olkiluoto 3, in Finland, and Flamanville, in France, will be in service by 2015, but Vogtle, and others that follow in the United States and elsewhere will come into operation after 2018. Nuclear plants will provide battery charging overnight. Wind farms and tidal generators can supply energy for hydrogen production when operating at times of low electricity demand.

9
The fallacy of renewables

The Green activists, almost to a man, are against the energy technologies which have served us well over the last 150 years, and advocate an energy policy of conservation and renewables. Most people would not argue with conservation which has been a driving force behind a lot of technical developments over the years.

Before the industrial revolution, which began at the end of the 18th century, renewables were the only energy forms we had. Wind propelled ships around the world and wind or water mills ground corn. Biomass (firewood) provided heat and transport was on horseback or in a horse-drawn vehicle. The global population was less than 1 billion and it was only because population began to increase as the result of advances in medicine that the inventions came which developed the basic infrastructure which has been with us for over 100 years.

Anyone alive today who was born before 1945 has during their lifetime seen the population of the world treble, and predictions are that it will have further increased by a half up to 2050. This has not happened so much in Europe, where birth rates are declining, so much as in Africa and Asia, which now boasts two countries with more than a billion people.

Another 3 billion people on the planet by 2050 represents a 25 percent decline in population growth compared with what has gone before. But we must not be complacent, these extra people will have to have food to eat, water to drink, and somewhere to live. Energy has to be supplied to meet these demands.

Those who advocate conservation and renewables don't seem to realise how much the world has changed since those halcyon days of yore. Even to go back 100 years to a time when much of the

9.1 Rio Parana, Brazil/Paraguay: completed in March 1991, with a total output of 14750 MW the station is no longer the world's largest but still has the largest hydro power output of over 90 TWh/year. (Photo courtesy of Itaipu Binacional)

present infrastructure had already been developed we have only to consider the gains in fuel efficiency which have occurred.

The fuel consumption of a 2-litre, gasoline-engined car in 1960 would have been about 11.3 litre/100 km. The equivalent vehicle in 2012 has a more powerful engine but a fuel consumption of only 7.0 litre/100 km; if instead it had a diesel engine the fuel consumption would be 5.4 litre/100 km, and the new hybrid electric cars can achieve about 2.0 litres/100 km.

Similarly a Boeing 747 with four large turbofan engines such as the Rolls-Royce RB211 can carry 450 people from London to New York in about 7 hours, whereas the propeller-engined aircraft of the 1950s carried only 100 people and took 16 hours for the trip. If this shows nothing else it is that the efficiency of transport has improved by leaps and bounds and the increased fuel consumption is due to more people flying and the more cars on the roads of the world.

Then if we look at the improvements in power generation there have been two developments. Improved efficiency and reduced emissions. When the first nuclear power plants appeared at the end of the 1950s it was a time when there was great concern over the

9.2 Sindouping, China: three Gorges is the largest hydro power station in the world with 26 x 700 MW generators in these two power plants and six more in an underground plant. (Photo courtesy of China Three Gorges Corporation)

efficiency of coal-fired power stations. The first units with reheat steam cycles had raised efficiency up to about 36%, but it would be another thirty years before materials technology had sufficiently advanced to support a supercritical steam cycle and raise the efficiency to around 45 percent.

The nuclear power stations were less efficient because they had lower steam conditions but they had no gaseous emissions and they could produce more fuel than they consumed. To exploit this property the spent fuel from the reactor had to be reprocessed to separate the uranium and plutonium from the fission products and produce mixed oxide fuel which could be returned to the reactor.

The first reactors were built in the United Kingdom, the Soviet Union, the United States and France. At that time these were the only four nuclear weapons states, so they had all of the technology at their disposal, uranium enrichment, fuel fabrication, reactor construction and operation, and spent fuel reprocessing. They all had the complete plan of nuclear power development mapped out in front of them. It was a new generation technology which had low fuel costs and zero emissions, and in the following ten years the designs of reactor evolved into much larger plants which were

TABLE 9.1: THE WORLD'S LARGEST HYDRO PLANTS

Country	River	Dam and power plant	Output MW
China	Yangtze	Three Gorges	18,200
Brazil/Paraguay	Parana	Itaipu	14,000
Venezuela	Caroni	Simon Bolivar	10,235
Brazil	Tocantins	Tucuruí	8,370
Canada	Churchill	Churchill Falls	6,988
United States	Columbia	Grand Coulee	6,809
China	Hongshui	Longtan	6,426
Russia	Yenisei	Krasnoyarskaya	6,000
Canada	La Grande	Robert Bourassa	5,616
Russia	Angara	Bratskaya	4,500
Russia	Angara	Ust Ilimskaya	4,320
Brazil	Sao Francisco	Paulo Afonso	4,279
China	Lancang	Xiaowan	4,200
China	Yellow	Laxiwa	4,200
Argentina/Paraguay	Parana	Apipe Yacerita	4,050
China	Dadu	Pubugou	3,500
China	Yalong	Ertan	3,500
Pakistan	Indus	Tarbela	3,478
Brazil	Parana	Ilha Solteira	3,444
Brazil	Sao Francisco	Xingó	3,162
Tajikistan	Vakhsh	Nurek	3,015
China	Wujian	Goupitan	3,000
Russia	Angara	Boguchany	3,000
Canada	La Grande	La Grande 4	2,779
United States	Black Creek	Bath County *	2,772
Canada	Peace	W. A. C. Bennett	2,730
United States	Columbia	Chief Joseph	2,620
Russia	Volga	Volzhskaya	2,572
United States	Niagara	Robert Moses	2,515
Mexico	Grijalva	Chicoasén	2,430
Romania/Serbia	Danubel	Iron Gates	2,428
Canada	La Grande	La Grande3	2,418
Iran	Karun	Masjid i Suleiman	2,400
China	Jinanquiao	Jinsha	2,400
India	Tehri	Bhagirathi	2,400
Turkey	Euphrates	Atatürk	2,400
Venezuela	Caroni	Carauchi	2,160
Brazil	Parana	Ibaltumbiara	2,082
United States	Colorado	Hoover	2,080
Mozambique	Zambezi	Cahora Bassa	2,075
United States	Columbia	The Dalles	2,038

* Bath County, VA is the world's largest pumped storage scheme

then under construction in several European countries, the United States and the Far East.

It was the 1973 war between Israel and its Arab neighbours that finally started to unravel the nuclear industry. There were then 98 reactors in commercial operation in fifteen countries. But the oil crisis which followed had political consequences which saw the defeat of centre-right governments in the UK, France, and the United States in the following Elections.

It was the United States and Canada that saw the beginning of the global anti-nuclear protest movement. It was not specifically anti-nuclear but rather of the opinion that we had quite enough energy and should use it more efficiently. Therefore there was no need to build any more large power stations. At the time the oldest nuclear plants were less than twenty years old.

The decades since 1945 had seen the doubling of electricity production every eight years. Coal- and oil-fired power stations had been built with high pressure reheat steam cycles for higher efficiency, and lower production cost. Oil-fired power generation was at its peak and the only large scale alternative to closing a 1500 MW oil-fired power station was to build either a coal-fired or nuclear power plant.

The first protests were over more mundane matters such as the use of lead tetraethyl as an anti-knock agent in gasoline, and the high sulphur content of some coals. Lead in the atmosphere would drive our children mad, with increased traffic in the urban environment; and high sulphur emissions would lead to acid rain. Lead in gasoline had been a long running campaign over many years, but acid rain was a more recent concern and was seen to be changing the acidity of lakes and rivers in Sweden which was downwind of many large coal-fired power stations in Denmark and the UK.

Unleaded gasoline was introduced in the United States in the mid seventies and at about the same time new coal-fired power plants in Germany and the United States were the first to be fitted with flue gas desulphurization (FGD). Forty years later unleaded gasoline is universal and FGD is a standard system for all coal-fired power, which has been accepted because the by-product gypsum can be sold to the building trade to make plaster board.

So here were protests which bore fruit, because they made a

9.3 Volgograd, Russia: the Volzhaskaya hydro station a few kilometres upstream of the city is the largest hydro station in Europe with a total installed capacity of 2572 MW. (Photo courtesy of Russhydro)

definite improvement to the environment. Following this, growing anti nuclear protest had by 1990 almost brought the industry in the United States and Europe to a standstill. Only France and Belgium had taken nuclear power to the point where it carried all the base load of electricity demand. The United States had abandoned closure of the fuel cycle and the only place to get fuel reprocessed was in either, France, the UK or Russia.

The only other possibilities for electricity production were the renewables, wind, solar, tidal and biomass, but not hydro. The protesters did not like hydro at all, at least large hydro which could interfere with fish migration and displace whole communities from flooded areas upstream.

Yet despite this hydro power is the only credible renewable energy form available to us. Hydro power has no fuel cost and no waste product. But it does modify landscape and flood large areas from which people have to move. Many dams on some of the largest rivers have not only been built for electricity generation, but for downstream flood protection, irrigation and improved navigation.

Historically hydro power has been large power stations on the major rivers. Examples are Aswan High Dam, 2100 MW, on the

9.4 Inga, Democratic Republic of Congo: Inga 1 station with canal leading to Inga 2 in foreground. Total capacity of 1770 MW in the two plants is sent to the copper mines in Shaba Province. (Photo Courtesy of SNEL)

Nile: Hoover Dam, 2080 MW, on the Colorado River, and Grand Coulee, 6500 MW on the Columbia River, both in the western United States; Itaipu, 12600 MW, on the Parana between Brazil and Paraguay and Tucuruí, 8370 MW, on the Tocantins River, which is a large southern tributary of the Amazon.

These power plants were not only built for power generation. Aswan and Grand Coulee were primarily for irrigation. Hoover Dam, completed in 1938 was one of the many projects of President Franklin D. Roosevelt to create employment after the great depression of the early 1930s. Itaipu was completed in 1991 and was the last of the major hydro sites in the south of Brazil, which at that time gained over 90% of its electricity from hydro stations on the Parana and its tributaries, also the Tocantins and Sao Francisco Rivers in the north. These plants spread electrification through the western provinces and up into the Amazon basin.

The world's largest hydro power countries are Brazil, Canada, China, Russia and the United States. Brazil has concentrated on the Parana and its major tributaries in the south of the country. The Sao Francisco is a large river in the northeast with two large schemes at Paulo Alfonso and Xingo, while Tucuruí is an 8370 MW plant on the Tocantins which is the last southern tributary of the Amazon before the Delta.

Another country which has large areas depending on hydro power, is Canada where three provinces, Quebec, Manitoba, and British Columbia have large hydro plants supplying most of the electricity demand. Ontario ran out of economically suitable sites in the 1950s and went first to coal-fired power and then to nuclear. Quebec and British Columbia have separate territories apart from the main hydro network. Quebec has a 675 MW nuclear plant on the south bank of the St Laurence. In British Columbia a 312 MW HVDC connexion was laid across to Vancouver Island, in 1967 followed by a second of 370 MW in 1978.

China has built some of its largest hydro plants in recent years as its economy has developed. The Three Gorges on the Yangtse River which is now the largest hydro power station in the world, with two surface power plants which have 26 x 700 MW turbines, the last of which went into operation in October 2008. The total capacity then was 18,200 MW. A third underground power plant with six more 700 MW sets was completed in 2011. There are two 50 MW turbines which supply the station auxiliaries. The total capacity at Three Gorges is therefore 22,500 MW.

Starting late on its big rivers China has adopted the largest hydro turbines and are continuing to use them on other hydro projects. For three Gorges there were two contracts for foreign consortia working with Chinese partners. One group was led by Alstom, with ABB, Kvaerner, and Harbin Motor, which supplied 14 turbines for all three power houses. The other group included Voith, General Electric, Siemens, and Oriental Motor.

The Gezhouba dam, 36 km downstream of Three Gorges, was in fact the pilot plant for the larger scheme. It was built at a point where the river widens from 300 m to more than 2 km and there are three islands. Three Gorges is not the largest hydro power producer, because there is a wide difference in the rate of flow during the year from a minimum 5000 m^3/s, between November and March, to a maximum flow of 30000 m^3/s in July. In the first full year of operation of 26 units, which was 2009, the stations produced 79.47 TWh, and in the following year 84.7 TWh. In the same year in South America, Itaipu on the Parana River generated

85.97 TWh in 2010.

Many of Russia's large hydro plants date from Soviet times to support industrial development. East of the Urals are three large Rivers; the Ob, Yenisei, and Lena, flowing north into the Arctic Ocean. In European Russia the Don flows into the Sea of Azov and the Volga into the Caspian. Damming of these rivers to improve navigation and generate power was undertaken during the Soviet time.

The largest power plant in European Russia is the 2600 MW Volzhskaya plant, the lowest and largest of a cascade of hydro plants descending the river between Kamas and Volgograd, and was completed in 1961. A 752 m wide, 44 m high concrete gravity dam was built across the river between 1951 and 1958. Supporting it is a 3250-metre long land-filled dam with a maximum height of 47 metres to enclose the reservoir. The 22 generators were installed between 1958 and December 1961. Of these, seventeen are rated at 115 MW, three at 120 MW and three at 125.5 MW.

Above this station are Rybinsk, at 364 MW, completed in 1941; Zhiguli. 2315 MW, completed 1957: Nizhny Novgorod, 560 MW, completed 1959; Saratov, 1360 MW, completed 1977; and Cheboksary, 1404 MW, completed 1988. In parallel with the early constructions a canal was built to connect with the Don so as to provide a connexion between the Caspian and the Black Sea.

Attention in the sixties turned to the big Siberian rivers, and in particular the Yenisei and its tributary the Angara. About 30 km upstream of Krasnoyarsk, on the Yenisei, a 104 m high concrete gravity dam was built and a 6000 MW power station which was completed in 1972. The energy is mainly supplied to the aluminium smelter at Krasnoyarsk.

There is still much hydro potential that can be developed, notably in Africa, Asia, and South America. Africa could potentially house the largest hydro plant in the world, Grand Inga, which at a planned 39,500 MW would be bigger than the combined outputs of Itaipu and Three Gorges.

The great diversity of energy demand across the continent is one of the problems. In total, electricity production in Africa is about 600 TWh of which South Africa with 5.5% of the population has a per capita consumption of 4500 kWh/year. Morocco, Tunisia,

Libya, Algeria and Egypt between them have 16.7% of the population and a significant industrial base with the oil industry and have a per capita consumption of 1000 kWh/year. This leaves the rest of Africa with 77.8% of the population and a per capita consumption of 250 kWh/year, which is less than half the average monthly consumption in Europe.

The Zaire River is the largest in Africa with a number of large tributaries. What makes it unusual is that the drop in the river level is concentrated near the mouth with a long series of rapids below Kinshasa. The entire river is estimated to have about 100,000 MW of hydro power potential, of which some 44% is concentrated near the mouth. The Inga falls a few kilometres below Kinshasa, where it drops 95 m at an average flow of 42,476 m^3/s.

Two power plants were built at Inga bypassing the falls. The first at 352 MW was completed in 1972 and the second, larger plant of 1400 MW was completed ten years later. The output of both plants went to supply the copper mines in the east of the country, but are now not running at full output for lack of maintenance.

The Democratic Republic of Congo has been racked by civil war as have many other countries in the region in much of the last 25 years and it is only now that the conflicts have died down and people are looking again at first, a third Inga power plant, but also Grand Inga which early in 2008 was the subject of a conference in London between bankers and utilities.

As currently planned there are two plants, Inga 3 at 3500 MW which is specifically to power an aluminium smelter, and Grand Inga at 39,000 MW. The cost is put at $80 billion and completion would be around 2025.

But where would the energy go? There is nowhere near the demand in Congo for the 372 TWh/year that the power station would produce. There are suggestions that high-voltage direct current lines could transmit it as far as South Africa and even to Europe and the Middle East. The lines would be several thousand kilometres long and could only be HVDC to avoid losing much of the output in reactive transmission losses.

Africa has the largest unexploited hydro power potential in the world, and half a billion people with no electricity supply and dependent on biomass (wood and animal dung) for their energy supply. There is the technology available to build the power plants and transmit their output over long distances but a lack of skilled engineers to maintain the systems in all but a few places.

But any new hydro development in Africa or elsewhere attracts Green protest because of environmental damage, the obstruction of fish migration and flooding behind the dam, and the consequent displacement of people and livestock. While it may be true that this has happened on some rivers, the fact is that historically, hydro power has been developed to improve irrigation and prevent the flooding of downstream communities.

These are real benefits and the development of hydro capacity would facilitate rural electrification in parts of Africa so that people could have a clean source of energy for heating, lighting and cooking and not have to rely on the use of crude biomass for their basic energy needs. The millions of people without electricity in rural areas of the tropics, and not just in Africa, are one of the biggest sources of pollution through their dependence on wood and dung for their energy.

As electricity demand and population grew in the wake of the second world war many countries reached the end of their hydro resources that could be economically exploited. To go to more remote locations with all the problems of access for heavy plant and equipment it was more economical to turn to thermal power. Some old hydro stations had closed and the equipment had been removed but the reason for their having been built is still there.

So while some of these hydro sites are being redeveloped there are also further opportunities for small plants to be developed in the deregulated environment. The advantage of hydro over all the other renewables is that it uses water which can be stored.

Large commercial rivers have been developed to facilitate barge traffic and arc a series of large ponds between weirs with locks to allow shipping to pass. In June 2011 it was reported that nearly one hundred applications had been filed for hydro developments on the Misissippi including new hydro plants at existing dams where there are none at present and the refurbishment and upgrading of existing power plants. Nineteen applications have been filed for sites on the upper river between Hastings, MN and Cairo, IL It is estimated that sites on the Misissippi and other rivers, if they were all to be developed could add 70,000 MW to the US grids.

There are also rivers which cannot be exploited for serious commercial reasons, particularly on parts of the Atlantic coast of Europe and the Pacific coast of North America because these are the breeding grounds for the Atlantic and Pacific salmon.

Construction of classic hydro plant takes several years. The river flow and its variation over the year determine the height of the dam and the size of the generating plant, and the first generator cannot be tested until the reservoir is sufficiently full. The only standardization is in the individual generators of the power station. but hydro power, has seen its role in electricity supply changed with the introduction of thermal and nuclear capacity. Its value is in its much greater flexibility of operation.

Small hydro schemes are a totally different system. In the early years of the 20th century many small sites were developed to provide electricity to the surrounding communities. There are others which would have been developed in earlier times but were overtaken by events. Many of these old sites are now being rebuilt and some others are being developed. These are relatively small schemes, generally run of river plants of less than 10 MW, but as green energy schemes earn a premium price for their output.

Deregulation of Electricity Supply in the UK created some opportunities for small hydro particularly in the southwest where a number of rivers flow off Dartmoor, and Exmoor. Several units have been installed to provide energy for companies or individual farmers and landowners either completely or with some export to the grid. But the big activity in small hydro is in Scotland.

First, Scottish & Southern built the 100 MW Glendoe hydro scheme in the western highlands above Loch Ness. This was completed in 2009. This is the only large scheme, but a number of smaller schemes are under construction, some being the refurbishment of old power plants built almost 100 years ago for the aluminium industry.

RWE Npower Renewables in the United Kingdom has built a number of small hydro sites. At present they have 15 plants with a total output 58.8 MW. Typical of these is Blantyre, some 15 km above Glasgow on the Clyde at an existing weir. A single Kaplan turbine generates some 575 kW and came into operation in 1995.

The company are constructing another four stations in the North of Scotland with a total of 10.75 MW. A fifth scheme at Romney Weir on the Thames near Windsor will have hydrodynamic turbines (Archimedes screws) installed in two bays of the weir which will produce about 1.4 GWh per year and will supply Windsor Castle. The scheme which could be similarly applied elsewhere on the river has a net capacity of 300 kW depending on the river flow which could be between 5 and 20 m^3/s.

Romney Lock with a drop of 1.46 m is not the deepest on the Thames; further upstream, Boulter's Lock at Maidenhead is the deepest at 2.39 m and above there, Marlow Lock is 2.16 m. Eleven years ago flooding around Maidenhead led to the construction of the Jubilee river an 11 km long flood relief channel from above the town to re-enter the river at Potts Island below Romney Weir, which was completed in 2002.

Pumped storage is a variation of hydro power which was introduced with the growth of nuclear power in the 1960s. With this arrangement water is pumped up from an existing lake to a reservoir constructed on the top of a nearby hill. The turbomachinery is a reversible pump turbine so that when water is pumped to the upper reservoir it acts as a pump driven by an electric motor. To drain the reservoir the pump is reversed and runs as a turbine driving a generator. One of the first was built at Blaenau Ffestiniog, in North Wales, another was built at Vianden, Luxemburg but the largest in Europe so far, with six 330 MW reversible pump turbines is also in North Wales at Dinorwig.

This was completed in 1984 and is entirely underground. The upper reservoir on Llyn Beris was recently enlarged. With the build up of nuclear power in a predominantly thermal system the station was built to carry the loss of two 660 MW sets dropping out simultaneously. All fourteen of the Advanced Gas-cooled Reactors had 660 MW turbogenerator sets, and besides these there were also six coal-fired units at Drax, and three oil-fired units at Littlebrook. So Dinorwig was built for system security with each of the six turbines able to run up to full load in 16 seconds.

This is the classic example of renewable energy coupled to a thermal power system. It was introduced as nuclear power plants came into service, in order to provide them with a night load. This

9.5 Blantyre, Scotland: one of more than 20 micro hydro sites in the UK. A 575 kW Kaplan turbine is in this plant at a weir on the river Clyde some 15 km above Glasgow. (Photo courtesy of RWE Npower Renewables)

the reactor keeps running and pumps water up to the top reservoir. Then in the following morning as electricity demand increases the reservoir starts to drain like a conventional hydro plant. It starts as soon as the head gate is lifted, and each unit can supply 330 MW of peak power for as long as it is needed.

This shows that any hydro power generator is a fast acting plant which can respond immediately to a loss of generation capacity. It is the only plant that can do this, provided that there is water in the reservoir. Hydro power has been available for a long time and many of the largest schemes are more than 50 years old. There are still a large number of rivers in Asia and Africa which have not been developed because there has not been the demand for the energy within reasonable range of the site.

Until deregulation 20 years ago a typical electric utility in an industrial country would have a few nuclear coal- and gas-fired plants, some hydro plants, and a few simple cycle gas turbines. Capacity would range from 25 MW for the gas turbines to say 1200 MW for the nuclear plant. Charges for electricity were for the overwhelming majority of customers affordable and always had been.

9.6 Dinorwig, UK: one of the six 330 MW Francis reversible pump turbines at Europe's largest pumped storage plant in north Wales. Built for system security in the 1980s. (Photo courtesy of International Power)

A little mentioned renewable concept, perhaps because of its relatively late development, is tidal power, specifically the marine currents. Marine Current Turbines Ltd. (MCT) of Bristol have been researching the concept since 1994. Their first commercially sized unit was installed in Northern Ireland at Strangford Lough. This 1.2 MW unit has been in operation since June 2008 and the maximum tidal flow into and out of the lough is 4.5 m/s. Performance has been higher than thought. The annual production is 6000 GWh and with an availability of 60%.

The advantage of tidal energy is its predictability. It comes from the rotation of the Moon around the Earth and of the Earth around the Sun. Because the tidal cycle is about 12.5 hours and occurs twice a day the times of high tide can be mathematically predicted long into the future.

The United Kingdom has been a centre of the marine current research and has a number of sites where marine turbines could be installed. The highest tidal range in the country is in the Severn estuary where it has been considered to build a tidal power station across the estuary similar to the French power plant in the Rance estuary which has been in operation since 1968.

But the Severn scheme has been finally rejected because of the environmental damage it would cause. The estuary is an important winter feeding ground for migrating Arctic birds. Avonmouth is the port for Bristol and the surrounding area and would be upstream of the dam site. The Severn bore caused by the tide running up an increasingly narrow river channel would also be lost when it is one of the largest in the world and popular with surfers.

Marine current turbines have none of these disadvantages. since the operate in the water on the ebb and flow currents which drive the tide by adjusting the pitch of the blades. Although the MCT prototypes have visible structure above water. This does not have to be and planned larger units are completely submerged.

The first orders for tidal power plants have been placed in the UK and Canada. Specifically, MCT have a contract from RWE Npower Renewables for five 2 MW Seagen units, similar to the one in Strangford lock off the north coast of Anglesey. There is a reef offshore known as the Skerries and the spring tidal current between the reef and the mainland is 3.2 m/s. The turbine axes will be at a depth of 28 m.

A second scheme is in the channel between Skye and the Scottish mainland at Kyle Rhea. Tidal currents peak at 3.5 m/s. The plan is to install four 2 MW Seagens working at a depth of 27 m. Both schemes will be progressed in parallel with the first units at both sites staring in 2014. Kyle Rhea will be complete in 2015 and Skerries in 2016.

The Canadian scheme is in the Bay of Fundy which has the highest tidal range in the world at about 15 m. The highest tide ever recorded, in October 1869 was 21.6 m which was a combination of high winds, low atmospheric pressure, and a spring tide. But on a typical day during the 12.4 hour tidal period 115 billion t of water flows in and out of the bay. Minas Basin Pulp and Power at Hantsport, NS is a manufacturer of liner board from recycled paper and board. MTC will supply a 3 MW submerged unit with three of the individual Seagen generators mounted on a horizontal frame which is totally submerged. The mean spring tidal current is 4.1 m/s and the turbines will work at a depth of 30 m. The project is due to be completed in 2015.

Two other British schemes are from the Norwegian company

Hammerfest Strøm who have designed a completely submerged unit which in a 300 kW prototype has been tested at 50 m depth in Kvalsund a large sea inlet east of Hammerfest. A larger unit rated 1 MW has been installed at the European Marine Energy Centre (EMEC) off Orkney. This unit is the prototype of the commercial units that will follow, Scottish and Southern Renewables have placed two contracts. One in Islay Sound with ten units, and another off the north coast of Scotland at Duncansby head, for 96 units.

The generators are designed for installation in water between 40 and 100 metres deep. However the units for the Sound of Islay will be approximately 30 metres at the axis of the turbines. The mean spring current is between 3.5 and 4 m/s at each site.

Of the other renewables biomass is perhaps the most difficult to reconcile with energy production. Biomass historically has been the waste materials of the forestry and pulp and paper industries and also some agricultural wastes. There was a time in the past when after the harvest the straw left in the fields would be burned off where it stood. This was hardly an environmentally friendly act for homes bordering these fields and for traffic on rural roads and was soon brought to a halt.

One of the first examples of bio energy at work was in Brazil in the aftermath of the 1973 oil crisis. Brazil was then an oil importer and the Government decided to plant more sugar cane, specifically to be distilled into alcohol to be mixed with gasoline. By the end of the decade, cars in Brazil were all fuelled by a mixture of 95% gasoline and 5% alcohol.

The other most widely used bio-energy is the traditional system of burning wood waste from saw mills and forestry operations and various agricultural wastes There are a number of power stations burning wood chips, municipal solid refuse, and farm wastes such as chicken litter, but these are relatively small. typically up to about 25 MW and located near to the source of the fuel though some with large industrial suppliers of waste can be much bigger.

This is what we normally understand by biomass. Since the fuel has to be taken by road to the power plant this limits the economic radius for fuel supply. So biomass has tended to be fuel for specific industries. However a number of coal-fired plants with

9.7 Artist's impression of Seagen Array for the Skerries off the north coast of Anglesey. One turbine set has been raised for inspection from the boat alongside. (Photo courtesy Marine Current Turbines Ltd)

limited operating hours because they are shutting down in 2015 are burning up to 10% biomass in the fuel.

The 44 MW Stevens Croft powerplant of E.ON (UK) at Lockerbie, in southern Scotland, is burning wood waste from sawmills and forestry operations in the area and at the end of 2008 was due to take the first consignment of coppice willow from local farmers who are growing it. In total it needs 480,000 tons of wood per year to produce its full output and the ultimate aim is to have approximately 20% of the fuel coming from coppiced willow.

This is E.ON's first biomass plant in the UK. They have a planning application out for a 30 MW plant at Blackburn Meadows, the site of an old coal-fired power station near Sheffield. Like the Scottish plant it will burn sawmill and other wood waste as well as coppiced willow, and miscanthus grown by nearby farmers.

Miscanthus (elephant grass) is best known as an ornamental grass plant which is available singly from garden centres across Europe. It is grown from rhyzomes which once planted sprout up year after year, and the biomass crop is derived from the temperate variety, Miscanthus Sinensis which can be grown commercially in Europe. Once planted it just has to be harvested every year with a

9.8 Hammerfest Strom marine current turbine is totally submerged and shown in this drawing as part of a much larger tide farm with the units set much closer together. (Photo courtesy of Hammerfest Strøm)

combine harvester. Any arable farmer can plant a few hectares as several have done.

Energy crops are not new. To go into the cultivation of energy crops there has to be the land available to grow them, and it must be profitable for the farmer to produce them. Clearly it was in the UK after 2000 but not now. Growth of the Chinese, and Indian, economies in particular has pushed up the price of food and raw materials which has resulted from higher prices in Europe and North America for the traditional crops such as wheat, barley and oil-seed rape.

Another factor which has increased crop prices is the use of large areas of wheat, particularly in the United States converted to produce ethanol which, as in Brazil, is aimed at reducing fuel imports; so grain exports have reduced as more is used for bio-ethanol production.

Gas created by anaerobic digestion of refuse in land-fill sites is another energy source and one of the largest schemes in Europe is in the UK near Solihull. Almost all municipal refuse in the UK goes to landfill and some sites have been organised to collect the gas released and use it to generate electricity. The landfill extended above ground in clay cells and sufficient gas is collected to drive a 3.5 MW gas turbine which supplies to Birmingham International Airport. Many smaller schemes use reciprocating gas engines.

In Europe a lot of the refuse is incinerated to generate steam for a turbine. Of course the Green movement don't like it and any plan to extend one of the existing 20 plants or build another in the UK meets with howls of protest from the Green Activists founded on ignorance. Up to fifty incinerators are planned in the UK which has one of the smallest components of incineration in its waste management policy of any country.

In September 2008 the British Government gave its consent to a 100 MW incinerator to be built at Runcorn, in northwest England. The fuel will be domestic and municipal refuse collected from Manchester, Liverpool and north Cheshire. The plant is being designed as a combined heat and power scheme which will supply process steam to the INEOS Chlor Vinyl chemical works.

Of all the non-fossil options for generating electricity biomass is potentially the largest with the use of waste material, from farms and forestry and solid municipal waste. In fact in the United States almost 4000 MW is generated by refuse incinerators and landfill gas sites around the country. In Denmark, every large town has a refuse incinerator which generates electricity and in winter time some of the steam turbines are connected into the local district heating systems.

In pulp mills the wood is digested in a solution of inorganic chemicals to separate cellulose fibres which leaves black liquor, an aqueous solution of lignin residues, hemicellulose, and the inorganic chemicals used in the process. The black liquor contains more than half the energy content of the wood in the digester. It is concentrated by evaporation and used as fuel which also recovers the inorganic chemicals (sodium hydroxide and sodium sulphide) used in the process.

There are about seven tons of black liquor produced for every ton of fibre. Black liquor recovery has been in use for 80 years and the best performing mills can recover about 99.5%. Some of the largest pulp and paper mills are in Sweden and Finland where the excess steam generation is fed to the local district heating schemes.

These are specific biomass processes to particular industries. But there are numerous substances in daily use which are difficult to dispose of where large volumes are involved; things such as vegetable oil and animal fat used in cooking in restaurants and industrial canteens. These are now being collected by companies which react them with alcohol to make bio-diesel which can be mixed as 5% of regular diesel fuel and used in ordinary diesel cars with no engine adjustments. Tests have been conducted with bio-diesel at higher concentration to optimize engine performance.

But how much do we need? Say there are 20 million cars in a country of which 20% are diesel and run on average 50,000 km/year at an average fuel consumption of 5.1 litres/100 km. So each vehicle in a year will consume 2550 litres of fuel. Biodiesel at 5% concentration increases fuel consumption by 2% so that the biodiesel car uses 2601 litres/year, of which 5% is biodiesel and the total population of diesel cars will burn 520,200 m^3/ year of bio-diesel.

There are certainly more than five million diesel powered cars in Europe and North America. In addition there are all the other commercial vehicles, particularly the buses and large articulated trucks, all of which burn diesel fuel at a much faster rate. If all of these vehicles must use a percentage of bio-diesel, from what do we produce it?

The world is now recovering from economic recession during which time there has been a significant reduction in greenhouse gas emissions. This could be due to any number of reasons: lower industrial production; higher unemployment; fewer people going out to restaurants, some of which will have closed.

But as economic activity increases, emissions will rise which means more vehicle movements and higher demand for biodiesel. Now if a country has declared that all diesel fuel must contain say 5% of bio diesel which may mean a total requirement per year of say 2 million m^3 of bio fuel, where will it come from? Are there enough restaurants and industrial canteens producing enough of the raw material or will there be more deforestation in southeast Asia to produce palm oil for export as green fuel.

In Europe the bright yellow flowers of oil-seed rape are a familiar site in early summer and although it is widely used for marga-

9.9 Lockerbie, Scotland: a 44MW steam plant burns wood waste from local forestry operation supplemented by coppiced willow grown by local farmers. (Photo courtesy of E.ON Renewables)

rine and cooking oil, it could also be grown for bio-diesel.

Therefore we must ask how we can have a bio-diesel fuel policy at a time when the world population is growing towards 9 billion with about 25% not having enough to eat; and with food crop yields lower than in the 1990s can bio-diesel be really considered a fuel for greener energy? Maybe not, but burning basic biomass in a sustainable plan to produce electricity and heat is certainly worth doing.

Solar energy is available during the daylight hours which is also when there is the maximum demand for energy. Furthermore the development of photovoltaic semiconductors has effectively brought the possibility of electricity supply down to individual consumers. This way, the visible impact of the panels is minimised by placing them on the roof of each building.

Photovoltaics respond to daylight and not just to the direct sunlight that they will have been set up to receive. There are already a number of large office buildings around the world with photovoltaic installations on the roof. Similarly a small array of cells can be installed on the south facing roof of a house. a 4 kW array would cover the daytime domestic load of a 4-bedroom house with

9.10 London, UK: photovoltic panels on roof of HSBC headquarters at Canary Wharf. The 77 kW unit is one of four that the bank has installed on premises in the UK. (Photo courtesy of HSBC)

the common electrical appliances. Any power not consumed on the premises would be sold back to the grid.

Utilities have now introduced buy back tariffs which enable domestic users to benefit from solar energy. Under this arrangement a price per kWh is defined. The householder will see the electricity meter turning backwards when solar electricity is used in the house and moving forward normally when the net electricity demand is higher than the solar output. When there is surplus power to sell then the utility pays at the buy-back rate. The contract in the UK at least, is normally for twenty-five years.

The International bank HSBC has installed four photovoltaic generating systems in the UK, of which the largest is on their headquarters at Canary Wharf in eastern London. Two others are at their Management Training Centre at Elstree, Hertfordshire, and another on the headquarters of their Internet Bank in Leeds.

Most photovoltaic installations cannot be seen from the ground, and even when they can as on the south facing roof of a private house they are only visible when facing the street. The inverters which convert the output to alternating current would be in the loft of the house and therefore invisible. They are completely silent in operation, which is on 365 days a year during daylight hours.

With minimum environmental impact being incorporated into the structure of buildings, photovoltaics can be considered a green renewable energy system. Not only that but solar energy is available at the time of day when electricity demand is at its maximum.

So now, from the sublime to the ridiculous. Wind was seen as the obvious natural energy source because mankind had used it before to propel ships and grind corn. But over the past 150 years these applications have been closed off because the wind was not reliable enough and could be positively dangerous to ships on the high seas.

The largest wind turbines currently available are 5 MW. But smaller units of about 3.5 MW have been built offshore. They have given a new term to the English language: Wind Farm - a group of wind generators on or off shore. Each one of these generators has to be mounted atop a column to catch the wind and have three blades about 60 m long. Out at sea it would be at least 100 m tall above the water line at high tide, and close in shore they might be standing in 30 m of water so allowing for foundations we might be looking at a tube of total lenth 150m and 10 m diameter at the base.

Out at sea it must withstand the strongest gales. The hub at the top of the column would be a structure housing the generator with sufficient space for maintenance crews to work. Outside it would be shaped to form a landing pad for a helicopter to bring workers and equipment to each site. Once unloaded it would go back on shore until summoned to return.

The world's largest offshore wind farm is the London Array the first phase of which is under construction off the Essex coast scheduled for completion in 2013. The project has 175 x 3.6 MW wind turbines by Siemens. The second phase will ultimately bring the output to 1000 MW. The initial project is about 25 km off the Essex coast and will cover a sea area of 100 km^2. There are two offshore substations to collect the energy and send it by cable to a new substation at Cleve Hill, near Graveney on the north Kent coast. In total there will be some 450 km of cable offshore.

The first foundations were installed in March 2011. In Figure 9.11 overleaf, the red pile will be hammered into the sea bed after which the yellow transition piece is installed on top of it. The two items are matched for length to accommodate variations in the seabed and ensure that the top of the transition piece is at the same level for all 175 units. As the foundations are installed so the cables are laid to connect the generators to the substations. Each transition piece has two tubes through which the cables are pulled to connect at the top platform. Eventually the circuit will be completed by making a separate cable connexion up to the generator. Components of the wind farm, the foundation modules, the masts and the generator assemblies to go on top are being shipped from Vlessingen, Netherlands. Construction started in earnest in 2010.

The wind farm is being built by Dong Energy, of Denmark, (50%) E.ON (UK) Renewables (30%), and Masdar, of Abu Dhabi (20%). Masdar Power is a company working on renewable energy projects in the Emirates. The Shams-1 project is a 100 MW photovoltaic solar plant for a new town, and their other project is a 30 MW on-shore wind farm. Abu Dhabi is aiming for 7% of electricity supply to be from renewables by 2020.

When the first phase of London Array is complete we will see if the availability is any better than the 35% which is typical for wind farms. It would produce 1931.8 GWh/year. The nuclear plant further up the coast at Sizewell, an 1100 MW PWR, contains less steel and copper than is required to build one of the London Array wind generators, and assuming there is a four week annual outage for refuelling and maintenance, would produce 8897 GWh/ year of base load power. Nuclear power stations are poor load-followers, but so then are wind farms. Two more reactors are planned at the site for service in 2020 and 2021.

Does this not show the fallacy of renewables: a nuclear plant with no emissions produces more than four times as much electricity as the wind farm, also with no emissions, which at best has only a third of the availability? Furthermore combined heat and power can only be supplied by the nuclear plant for industrial process heat supply or district heating.

The biggest problem with wind is the enormous quantities of materials required for a relatively small output. A tube 100 m long by 10 m diameter, just consider the energy needed to make it, and then how is it transported to site? Each unit must make the same

9.11 Installation of one of 175 turbine foundations for the London Array. The red pile and yellow foundation are designed to a give common height above the water line. (Photo courtesy of London Array)

journey out to sea and can it be assembled out there or must the complete mast and generator package be assembled on shore and transported vertically to the site and lifted on to its foundation?

In the United States all operating wind farms are on shore with California and Texas with the most installations. Wind generators are than offshore and atypical wind farm would have more than one hundred units rated at 1500 kW. Iberdrola Renewables own and operate the Shiloh wind farm which produces 375 Gwh which is equivalent to an availability of 28.5% over a full year.

So one unit of the wind farm contains more steel than would be required for major components of the nuclear plant, just for the column to support a single 3.7 MW generator some 100 m above the water line at high tide.

So there is another environmental impact of both plants and that is the impact of the mining of the production of the materials and components and their transport to and assembly on the final site. The wind farm requires about 120 times as much steel as for the single nuclear plant to produce a quarter of the output, and at a lower availability.

Not only that but the 450 km of submarine cable required to

9.12 Solano County, CA: some of 100 GE 1.5 MW wind generators in the first phase of the project completed in 2006. (Photo courtesy of Iberdrola Renewables US)

bring London Array power ashore compares with perhaps a 20 km overhead line to connect the nuclear plant, or indeed any other power plant on land, to the nearest grid substation.

At the height of anti-nuclear protests it was often said that the energy used to construct a nuclear plant was so great that it could never be recovered from its output over its lifetime. There is a much stronger case that can be put in the case of a large wind farm.

Onshore and offshore wind cannot therefore be considered a green energy system by any stretch of the imagination, because of the enormous amount of materials required even for one 3.7 MW wind generator offshore, and the energy cost of installation and assembly, as compared with a nuclear plant of 1100 MW.

But the other aspect of wind is its susceptibility to a wide range of wind speeds. During the warm summer of the American mid west it was reported that over 4000 MW of installed wind capacity could only produce 180 MW of power. In Europe as the tail end of a Caribbean hurricane blows across the Atlantic stronger winds are felt on land and wind farms have been shut down to avoid local overload of the grid. This happens quite often and compensation paid to the wind farm has to be carried by the consumers in higher energy costs.

Other renewables, however are more manageable. Biomass is no different from any other thermal power station. It produces steam for a turbine and although its efficiency may be relatively low, it is producing energy from waste materials including household refuse which would otherwise go to land fill.

Tidal current generators are a new technology which depends on a phenomenon which can be predicted a long time ahead. It does not run continuously but at variable times twice a day during the month. In this regard it is like wind energy in that it can present its peak output at a time when the demand for electricity is at a minimum.

Solar, or rather photovoltaics work during daylight hours and while there are considerable variations between summer and winter in northern latitudes, it can nevertheless be a worthwhile energy source for those who install solar panels.

But it can also be used in conjunction with another energy system. A set of mirrors could focus the sun onto a boiler which would supplement the steam supply to, say, a combined cycle. In the day time the steam turbine has a higher output than at night when it is running off the heat recovery boiler alone.

So there are various ways in which renewables can help with the new energy needs. High tide at night coupled with a strong wind provides energy to electrolyse water and produce hydrogen fuel for cars.

10 What is the future of electricity?

Anybody who buys this book immediately after publication will not live to see the year 2100, but they will certainly see a change of lifestyle as surely as has occurred in past centuries. The developments at the end of the Second World War have revolutionized transport where large gas-turbine powered aircraft have displaced the big ocean liners for global passenger transport and widespread car ownership has taken over from the horse and cart and the railways which were commonplace at the start of the 20th century.

In the 21st century then possible changes will be how we build and heat our houses, and what sort of cars people drive. Above all else what will be the population of the world in 2100, because that will have a bearing on much more that happens after 2050. Will there be mass migration to escape famine and drought?

Should this happen it will surely be from the tropics to the higher latitudes which has already given rise to public criticism in the receiving countries, where large populations of different race and religion seek to impose their way of life and not integrate with general society.

One consequence of this is the greater economic activity it brings, which of course pushes up the demand for energy. Ten years ago the average per capita electricity consumption was about 5700 kWh/year, in Europe. Rising population and greater use of new electrical items such as computers and microwave cookers and the fact that many more people work from home for some or all of the time has now pushed it up to more than 6000 kWh/year despite the increasing use of low-energy lights.

It could be said that the nineteenth century at the end brought basic infrastructure; electricity, gas, mains water and tarmac roads, and railways for long-distance travel. In the twentieth century we focussed on speed. The gas turbine is not only a power plant for an aircraft but also an electricity generator. The Concorde supersonic airliner was the ultimate flying machine with a flight time from London to New York of about 3.5 hours. But coming into service just after the first major oil crisis it was limited in use and it was the large turbo-fan aircraft which brought the world cheap travel to distant countries.

If there is a big development theme at the start of the twenty-first century it is the environment. How do we generate energy in a clean manner with fewer emissions and at higher efficiency. Some developments have improved the efficiency of engines which has been concealed by the increasing population and much greater car ownership. Coal-fired power plants are becoming more expensive to build, and public hostility to nuclear power following the Three Mile Island and Chernobyl accidents almost brought the industry to a halt in Europe and North America in the 1990s.

The anti-nuclear fanatics who had almost brought this about were for an energy policy based on conservation and renewables. Conservation has been a guiding principle in much of engineering design, particularly in engines, which if they produced more energy from less fuel would cost less to operate.

During the 1990s, global warming came to the fore as an issue which would need to be addressed lest it brought chaos to global agriculture and human lifespan. How many animals, birds and insects would be able to cope with a supposed 4 deg C increase in global temperature by 2050?

Whether global warming is a natural occurrence or man-made it has given an environmental push to technology in this century. The big benefit of this is the revival of nuclear energy in answer to demands of government for low emissions in power generation, and the inadequate performance of wind farms among renewable systems all but two of which cannot operate continuously.

When the first nuclear plants were built in the 1950s there was much public enthusiasm for the new technology. It was thought to be so cheap that electricity could be given away, but that was the first misconception. The high energy density of the fuel meant that the relative fuel cost was low; but the need to contain securely the radioactivity increased the capital cost. But this was a challenge to the designers and the result was the design of large reactors over 1000 MW which could produce a large output to offset it. Today a 1300 MW reactor, even with its relatively low steam conditions, has the lowest production cost per kWh of any thermal generating system, and the lowest emissions.

The other aspect of nuclear history is that it was the smaller power reactors which had the accidents. Three Mile Island was only 800 MW, Chernobyl was 1000 MW, and Fukushima Daiichi were 330 and 760 MW, which was the only Pacific coast nuclear plant to be damaged by the March 2011 tsunami. Serious as these accidents have been the latest has not halted nuclear development. but have slowed it while utilities and governments absorb all the lessons of the incident. Chernobyl did the greatest damage globally because it blew up the reactor and spread radioactivity over northern Europe. It resulted in public hostility engendered by the Green activists in several countries. Twenty years later a nuclear revival is gathering pace.

Power generation is not the only source of emissions but is one of the most controllable. These are visible emissions; smoke from stacks and steam from cooling towers. However the big invisible energy wastage is of heat from buildings. Some of this has been curbed by individual householders in fitting double glazing and insulation in their loft space. In many English towns one can see whole streets of houses built since 1950 in which every house has had all the original wooden framed windows replaced with the characteristic white upvc frames of double glazed windows.

While windows are easily replaced and loft insulation can be fitted to any depth, heat loss through walls cannot be easily curbed. Cavity wall construction of houses began in the early twentieth century and can be back fitted by spraying in small balls of insulation or foam. Modern building standards require walls to be insulated and for this compressed wool or straw is often used. Double glazing and loft insulation are now standard.

How will we heat our homes in future when the gas runs out or becomes too expensive? Electric heating by night storage heaters is practical and economic in a well insulated house and another possibility is to use a ground-sourced heat pump which can be likened to a refrigerator cooling your garden which is being heated by the sun. The output would be applied to the existing radiator header and hot water tanks.

In both cases the heating system has no emissions. These have been promoted in Scotland where they have enjoyed a strong take up of heat pumps and also photovoltaic solar panels in rural areas. Elsewhere oil is still being used for space heating where no gas supply is available but insulation and electric heating are practical alternatives.

As buildings are replaced the designs will minimise heat loss but given the number and age of all the buildings in a country it will take a long time to achieve a substantial reduction in heat loss. However a well insulated house will have a lower heating bill because it requires less fuel to maintain the desired temperature inside.

The other changes are not so much in curbing energy waste as changing energy use. Car ownership has expanded to such a degree that in Europe and North America the car is a personal possession. How then do we cut vehicle emissions given the extent to which so much goods traffic is taken by road, in addition to car use for personal transport to and from one's place of work?

The ultimate determinant will be the price of a barrel of oil which peaked at about $150/bbl in the summer of 2008. But this cannot be looked at in isolation because the cost of fuel is a factor in the cost of food and all the other things that we buy. In 2011 there is a major economic crisis in Europe which threatens another recession.

A Ship docks at a port with a cargo of containers. Each of the containers is placed on a truck which carries it to the company that has bought the contents, which may be only 20 km from the dock or 200 km. The full container is dropped at the destination and an empty one picked up which might be taken to another company who will fill it and send it back to the port.

Freight containers, when they were first developed, had been designed to be carried on a train or individually on a large truck. They quickly became the preferred means of transporting a variety of products which were traditionally shipped in small quantities.

For example a factory on Taiwan making computers boxes them and stacks the boxes in a container. This is then taken to the port where it is put on a ship for, say, Rotterdam. The truck brings back an empty container which is then filled and is put on a ship going to Sydney. The bulk of goods going by sea are now in containers and it is mainly bulk commodities such a coal, oil, LNG and other minerals which are carried in tankers and bulk carriers.

Freight shipping has been greatly simplified, but the nuclear cargo ship has yet to appear. Is this what so many of the small reactors under development are directed to? The reactors use highly enriched fuel which means that they have only to be refuelled, say, once every five years. This would greatly reduce oil consumption for sea transport if these were first applied to container ships and bulk carriers.

The problems of changing fuel come on land. Cars are such an important means of transport that people may not want to give them up for public transport. The problem is the short range of the battery driven electric cars. But is this the right technology?

It is notable that in the United States the alternative cars are either the hybrid cars with a gasoline engine in a series parallel arrangement with the electric drive motors; or the electric car powered by a hydrogen fuel cell, which generates electricity on the move for the drive motors. Vehicles of both types have been produced by the global motor industry and now being tested by customers in the United States. The hybrid gasoline-engined cars are already in production and several millions have been sold.

What this tells us is that road vehicles can be developed which can be used in exactly the same way as gasoline-engined cars at the present time. Several countries plan to install hydrogen filling stations which will be similar to present filling stations. Some of the first installations may be put on large forecourts, separate from gasoline services as commercial diesel pumps have been in the past. It would seem that the cars of the future will be hybrids and have similar performance as that of present vehicles.

But they still have to be fuelled. It is no problem for the hybrid gasoline-engined vehicles. A vigorous marketing campaign would see more of them on the road, putting the emphasis on low fuel consumption with high performance. If in five years time for a country with 20 million cars 80% were hybrids, that would be a 90% reduction in demand for gasoline. With hydrogen powered hybrids expected to enter the market in about 2015 we could see a further reduction in demand for gasoline and diesel.

The consequence of this will eventually be that in summer time cities will seem to feel more humid because the combustion product of hydrogen is water. Hydrogen is not difficult to produce since it can be produced in sewage works as in California, or by electrolysis of water. How much hydrogen could be produced by a 25 MW offshore wind farm which would not be connected to the grid but to a hydrogen plant on shore? Given that the wind farm has lower availability and that hydrogen can be stored until it is used, this may be the easiest way to produce it.

A 400 MW combined cycle runs in the day time and is shut down at night or only runs at part load. It therefore could maintain its full output on weekday nights and feed a number of hydrogen plants around the country. Furthermore it would be running at a higher efficiency than otherwise.

If we take coal out of mainstream power production we have to accept nuclear energy. Of course if the Green movement had not been able to convince government of their false arguments during the last decades of the twentieth century, we would not have now three governments at least refusing to install this clean emission-free energy system. Perhaps there would be more than 442 nuclear plants in operation and the global warming scare would not have yet been heard.

There is another issue which will reward emission-free power generation: the carbon tax. This is already being introduced and gives a value to a ton of carbon dioxide emitted. In its simplest form the tax is paid by the operators of power plants and industrial boiler plant burning coal, oil, or natural gas. It does not cover all carbon emissions as the few countries who have introduced such taxes have not applied it to gasoline and diesel fuels for transport.

The idea is that carbon taxes should be offset by reductions in corporation taxes. This would have neutral effect on the overall tax burden of a company. If not the company would have to recover the tax from its customers through higher prices for its products. This would apply as much to the electricity producers as to the industrial combined heat and power operators.

The carbon tax must be set at such a level that it encourages the use of cleaner energy systems. Various existing measures to control emissions are already having an effect. For example the lack of a large-scale carbon capture system for the total output of a coal-fired power station means that there are very few being built and in some countries there is strong opposition to building any more.

A large number of coal fired power stations in Europe will close down in 2015 if not before. They are unlikely to be replaced, not only because they will be less efficient, and therefore will burn more coal to produce the same amount of electricity. But they will also have to pay the carbon tax. Coal is effectively being priced out of the power market.

The fallacy of renewables is that their intermittent output means that there must be stand-by plant to take over when the output is low or non-existent. This is why there has been development of gas turbines which are not only more efficient but can ramp up their output much faster than before. While these can certainly come into play when the wind stops blowing or the sun goes down, it is not unknown to shut down wind farms in anticipation of a sudden demand for energy as for example, a television peak.

Then again, is biomass all that it is claimed to be? Bio diesel from reprocessed cooking oil and animal fat is used at about 5% in regular diesel. But is this sustainable? Does the country that can reprocess it import more of the raw materials from countries that do not have the technology, so they can drive diesel vehicles with 100% bio fuel while their supplier countries cannot?

The other side of biomass is whether a small power plant burning locally sourced waste wood and energy crops should be built, or a large coal-fired plant is converted to burn wood chips imported from half way around the world while the supplier continues to burn coal or natural gas. The value of biomass is that it is a way of producing useful energy from waste materials from farm and forest and restaurants which were not all easily disposed of in the past.

In the twentieth century increased demand for electricity resulted in larger power plants. Steam turbines of between 300 and 1000 MW were installed in coal and oil-fired plants because these would be simpler to erect, and commonality of design simplified maintenance. The move to larger sizes ensured that costs were kept down and this was something that the nuclear industry followed to avoid competitive disadvantage.

Electricité de France in particular had a long experience in building thermal power plants to a standard design. The principle was carried forward to their nuclear plants. Initially they installed 32 nominally 900 MW units between 1977 and 1985. These were followed by 22 of the 1300 MW design between 1985 and 1997. Two examples each of the upgraded N4 design at 1450 MW were installed at Chooz on the Belgian frontier, and at Civaux, south of Poitiers between 1995 and 2000.

The N4 design is the basis of the 1600 MW EPR which was jointly developed between France and Germany. The first EPR is nearing completion at Olkiluoto, Finland, with two more in France at Flamanville and Penly, under construction, and four planned in the UK at Hinckley Point and Sizewell. With plants planned or under construction in Turkey, Abu Dhabi, India, China, Korea, Taiwan and Japan, the industry would appear to be in good health.

The next ten years should see a large number of nuclear power plants built, the majority of them pressurized water reactors of between 1000 and 1600 MW capacity. But the period should also see some of the smaller reactor designs come into operation. These are essential if we are to produce more emission-free energy for base load power around the world.

But we also see the beginning of a family of small reactors with designs for water-, gas-, or liquid-metal cooling and which are aimed at specific applications. Furthermore, all are designed for installation below ground. This reduces construction costs since it does not require the high strength containment structures designed to withstand a crashing aircraft.

It is inevitable that our future life-style will be more electricity dependent. What is the likelihood that nuclear power will expand into industrial energy supply and district heating, since these are applications which use smaller generating sets of 300 MW or less. What happens if these companies are forced to add carbon capture to their plants. The small reactors of 300 MW or less are designed for installation underground, with just the steam turbine and the control room above ground.

Industrial combined heat and power got going in North America in 1980 and in the rest of the world some ten years later so that there are a number of plants which will be over thirty years old in 2020. The majority of the plants are combined cycles with, for example, three 120 MW generators and two heat recovery boilers which eventually will have to fit carbon sequestration.

First would it make sense to make the investment on a 30-year old plant as a condition of it staying in operation? There might be if there was a use for the recovered carbon dioxide. But with no market for the carbon dioxide what must they do? They can only recover the cost by adding to the cost of their products.

One process which can use the carbon dioxide recovered from a company's power plant is fertiliser where the carbon dioxide is reacted with ammonia to make urea. Given that many combined heat and power plants are gas fired the amount of carbon dioxide is relatively low compared with a coal-fired boiler.

All news of carbon sequestration is of experimental installations on selected coal-fired plants so the possibility is that a practical system for a full sized power plant of 800 MW is unlikely to be available much before 2020. But who will install it if either it adds excessively to the auxiliary load, of the station so that there is less electricity that can be sold, or it reduces the thermal efficiency.

Carbon sequestration is another green idea that assumed that a purification process in the natural gas industry could be applied to power generation without considering the much greater volumes of gas involved and what one should do with it.

What seems to be more likely is that coal will eventually be taken out of power generation. Certainly to take carbon dioxide out of the flue gases, compress it and pipe it hundreds of kilometres to an offshore oil field for enhanced recovery, will push up the cost of energy and moreover reduce the efficiency of production.

The fact is that in the previous century the global energy supply was produced inefficiently. The many nationalized electric utilities saw it as their job to supply electricity as cheaply as possible and nothing else. It was only when these industries were deregulated that combined heat and power really took off. At the same time the use of natural gas for power generation brought with it a higher efficiency of production.

Combined heat and power brought benefit to industry while the steadily increasing efficiency of the combined cycle has resulted in a greater output of cleaner energy with lower emissions.

However public opinion is starting to wonder why the price of electricity is rising. Is it because renewable subsidies are falling and their governments are still in thrall to the Green activists who forced them to include renewable energy systems in the plant mix?

If coal is not generating electricity in future what do we put in place of an old 1500 MW station? We can replace it with three 500 MW blocks of single shaft combined cycle at 60% efficiency, or a 1500 MW nuclear reactor with no emissions. If we are to have renewables then they must be predictable. That means solar which can be mounted on individual houses and office buildings to give an effective reduction in energy costs; marine current turbines which generate twice a day at predictable times; or biomass which, if sustainable, is burning waste material and household refuse.

For transport can we really have battery driven electric cars? Those with hydrogen fuel cells to charge batteries on the move for the electric drive motors are filled like the present gasoline-engine cars. Public opinion will want cars which are easy to fill up and have a similar range to present day vehicles.

Hybrid cars would allow existing assets which can generate at night to produce the fuel, which will beneficial to the performance of the utilities through its higher availability. We surely cannot afford to keep building large power stations just to provide the power to charge millions of electric car batteries throughout the night.

So the true green energy systems are Nuclear, Hydro, Natural gas by its applications of high efficiency and the three predictable renewables (solar, marine currents, and sustainable biomass) to reduce our energy costs and provide an essential, environmentally friendly fuel for future transport.

Index